HOUSEHOLD VULNERABILITY AND RESILIENCE TO ECONOMIC SHOCKS

Ashgate Economic Geography Series

Series Editors:

Michael Taylor, University of Birmingham, UK
Peter Nijkamp, VU University Amsterdam, The Netherlands
Jessie Poon, University at Buffalo-SUNY, USA

Innovative and stimulating, this series enlivens the field of economic geography and regional development, providing key volumes for academic use across a variety of disciplines. Exploring a broad range of interrelated topics, the series enhances our understanding of the dynamics of modern economies in developed and developing countries, as well as the dynamics of transition economies. It embraces both cutting edge research monographs and strongly themed edited volumes, thus offering significant added value to the field and to the individual topics addressed.

Other titles in the series include:

Global Companies, Local Innovations
Why the Engineering Aspects of Innovation Making Require Co-location
Yasuyuki Motoyama

Economic Spaces of Pastoral Production and Commodity Systems
Markets and Livelihoods
Edited by Jörg Gertel and Richard Le Heron

Towns in a Rural World
Edited by Teresa de Noronha Vaz, Eveline van Leeuwen and Peter Nijkamp

Household Vulnerability and Resilience to Economic Shocks

Findings from Melanesia

Edited by

SIMON FEENY
RMIT University, Melbourne, Australia

LONDON AND NEW YORK

First published 2014 by Ashgate Publishing

Published 2016 by Routledge
2 Park Square, Milton Park, Abingdon, Oxfordshire OX14 4RN
711 Third Avenue, New York, NY 10017, USA

First issued in paperback 2016

Routledge is an imprint of the Taylor & Francis Group, an informa business

Copyright © Simon Feeny 2014

Simon Feeny has asserted his right under the Copyright, Designs and Patents Act, 1988, to be identified as the editor of this work.

All rights reserved. No part of this book may be reprinted or reproduced or utilised in any form or by any electronic, mechanical, or other means, now known or hereafter invented, including photocopying and recording, or in any information storage or retrieval system, without permission in writing from the publishers.

Notice:
Product or corporate names may be trademarks or registered trademarks, and are used only for identification and explanation without intent to infringe.

British Library Cataloguing in Publication Data
A catalogue record for this book is available from the British Library

The Library of Congress has cataloged the printed edition as follows:
Feeny, Simon.
Household vulnerability and resilience to economic shocks : findings from Melanesia / by Simon Feeny.
pages cm. -- (Ashgate economic geography series)
Includes bibliographical references and index.
ISBN 978-1-4724-1919-4 (hardback : alk. paper)
1. Melanesia--Economic conditions. 2. Households--Melanesia. 3. Financial crises--Melanesia. I. Title.
HC681.45.F44 2014
330.995--dc23

2013029929

ISBN 13: 978-1-138-27669-7 (pbk)
ISBN 13: 978-1-4724-1919-4 (hbk)

Contents

Contents

List of Figures and Tables

Figures

Tables

Notes on Contributors

Matthew Clarke is Professor and Head of the School of Humanities and Social Sciences, Deakin University. His research interests are in religion and development and aid effectiveness and development in the Pacific. Professor Clarke also undertakes regular evaluations of community development programs in the Asia-Pacific region.

Jaclyn Donahue has worked in the not-for-profit sector since 2005 in the varied roles of community and volunteer liaison, programme manager, editor and researcher. Her research interests include the impacts of labour standards and access to food and nutrition on gender equality and poverty. She earned a Master's degree from the University of Denver's Josef Korbel School of International Studies.

Kate Eccles is a volunteer with Oxfam Australia and is also currently undertaking a Post Graduate Diploma in Economics at La Trobe University, Melbourne. Kate has experience in communications and has worked as a radio show producer and a media analyst. Her research interests include economics, gender, and development in the Pacific Islands.

Simon Feeny is an Associate Professor in development economics at RMIT University, Melbourne, Australia. He has published widely on international development issues. His research interests include the allocation and effectiveness of foreign aid, the Millennium Development Goals, poverty and human well-being and public sector economics. His research has focused on the Pacific region.

Lachlan McDonald is a development economist working with both Oxfam Australia and The Fred Hollows Foundation. Prior to this he gained several years of experience in central banking, both in Australia and Vanuatu. He has also worked as consultant with the Vanuatu Utility Regulatory Authority on a rural electrification project. Lachlan has a Bachelor of Commerce and Bachelor of Economics, with honours, from Monash University and a Masters of International Relations from the University of Melbourne. He is currently a PhD candidate at RMIT University, Melbourne.

May Miller-Dawkins is Research Director of CORE Collaboration Group in New York. Her current research projects focus on private-hybrid global governance and corporate accountability. She was previously Research Manager at Oxfam

Australia where she led research into economic resilience, human impacts of the global economic crisis and gender justice in countries across Asia and the Pacific Islands.

Manoranjan Mohanty is an Associate Professor with the Development Studies Programme in the Faculty of Business and Economics, the University of the South Pacific, Fiji. His areas of interest include urbanization, migration, NGOs and civil society, social development, the environment and climate change, and sustainable development.

Vijay Naidu is Professor and Director of Development Studies and Head of School of Government, Development and International Affairs at the University of the South Pacific, Fiji. He has recently completed a study on inter-ethnic relations and discrimination in Fiji for Minority Rights Group (MRG) International and is currently researching urbanization and informal settlements in Fiji as well as reflecting on the post-2015 development agenda.

Alberto Posso is a Senior Lecturer at RMIT University, Melbourne, Australia. He holds a PhD in Economics from the Australian National University in Canberra, Australia. His core research area is in labour and trade economics with a focus on the developing world. Alberto has predominantly authored studies on the development experience of Latin America, East Asia, and the Pacific.

Foreword

The Pacific island economies are characterized by their smallness, openness and dualistic nature. Their openness means that they are affected by events in the world economy which are outside of their control. Recent examples include the global financial crisis which resulted in the downturn in the world economy and preceding this crisis, the global commodity price increases which resulted in increases in fuel and food prices. These events have had a big impact on livelihoods in the Pacific. Their dualistic nature means that some of the population resides in rural areas. In Melanesia in particular, the majority of the population resides in rural areas and livelihoods are sustained by subsistence living which remains a buffer towards sudden shocks. However, the rate of urban migration in Melanesia is rapid and this has resulted in increasing poverty and hardship in urban areas as well as exerting huge pressure on urban resources.

This book is timely and salutary. The majority of studies on the above issues in the region are analysed from a macroeconomic approach and very few at a household level. For the first time in the Solomon Islands and Vanuatu, an extensive survey was carried out to determine the vulnerability of households to unexpected shocks. The book provides a useful insight into how households in these two countries respond to such shocks, especially with regard to increases in the prices of food and fuel and the impacts of the global financial crisis. The research focuses on two main household groups. The first is urban settlements and the second, rural households. Rural households are further divided into those that depend mostly on agriculture; households with close proximity to the urban centres and finally those that are very remote from their capital cities. This approach explores the different dynamics of vulnerability to shocks and it is highly commended.

The researchers have taken a multidimensional approach to analyse vulnerability. Their findings reveal that poor urban households were among the worst affected by recent increases in food and fuel prices and this adversely affected their standards of living. Their access to land resources is limited compared to rural households, restricting their ability to adjust during difficult times. Rural households that are closer to urban centres are less vulnerable, especially households that have land available and are partly engaged in the cash economy. Rural households in remote areas can be resilient since they can turn to their gardens and this demonstrates the importance of the customary economy in these countries. The findings for urban areas are important and should be a cause of concern, particularly with the high rates of urban migration in these two countries despite some movements back to rural areas. Without addressing these issues, poverty and hardship will continue to increase which are already causes of social disorder. As in other developing countries, poverty and hardship in the urban areas can only be improved by meaningful economic growth that translates into better employment opportunities. The book's findings further highlight how women are adversely affected by shocks and how their workload increases as a result. This is an important revelation and points to the fact that more gender based policy measures should be adopted to address this problem in these countries.

Overall these are important findings which are startling. I highly recommend that Governments and development practitioners read the book and incorporate its findings into their development thinking. For Governments in particular, the findings are important for their development planning and to prioritize their national developments. They need to address pressing issues concerning urban centres and to provide better links and investments that reach rural households. This should be via better infrastructure and sustainable economic projects to improve household resilience. Furthermore, Governments must address the inequities that arise from shocks, especially with regard to women taking a disproportionate share of the adjustment costs. This is in addition to their important mandate to ensure that macroeconomic stability is achieved at all times. Finally, I would also recommend that this type of study is replicated in many other countries.

Odo Tevi
Former Governor
Reserve Bank of Vanuatu
May 2013

Preface

In 2009, I was seconded to Oxfam Australia from my university on an industry placement scheme. At this time, most countries around the world were suffering the impacts of major global economic events. The prices of both food and fuel had increased dramatically in 2007 and 2008 and these price hikes were closely followed by a Global Economic Crisis (GEC). In 2009 global output contracted for the first time since the Second World War. The crisis initiated numerous conferences and generated a plethora of journal articles, policy documents and commentaries on its impacts in developing countries in Latin America, Africa and Asia. Yet, evidence of how recent these economic shocks were affecting Pacific countries was very limited.

During my secondment, with the help of Oxfam staff, I therefore focused my efforts on evaluating the impact of the GEC on countries in the Pacific region. Through the collection and analysis of macroeconomic data it was possible to examine how the crisis was affecting trade flows, government revenues, remittances and tourism receipts, for example. The study was able to capture the different mechanisms through which the crisis was being transmitted to the region and how the impacts varied across countries. Yet there were important questions that the study was unable to answer. How was the GEC impacting on households and communities? Were there gendered impacts? Had these recent economic shocks increased the incidence of poverty? There was a dearth of information to answer these questions or to assess how households in the Pacific respond to shocks more generally. This is particularly surprising given the region's vulnerability to natural disasters and climate change.

Sometimes the impact of the GEC on income poverty is estimated, based upon the impact of the crisis on economic growth. However, such an exercise is not appropriate for Pacific countries. Firstly, income poverty is not a particularly useful measure of well-being in countries where a large proportion of the population lives a semi-subsistence lifestyle in rural areas. Secondly, the relationship between economic growth and poverty in the Pacific is weak with increases in the former, not always leading to falls in the latter. Predicting impacts on poverty from falls in economic growth is therefore problematic.

The absence of information on how Pacific households are affected by global economic shocks and how they respond during difficult times provided the motivation for the research project underpinning this book. It focuses on two Melanesian countries: the Solomon Islands and Vanuatu. As demonstrated by the book, households and communities in Melanesia live distinct lives and lessons from the food, fuel and economic crises from other parts of the world are unlikely to apply. It is hoped that the insights provided by this book will spur further research into the increasingly important issues of vulnerability and resilience to shocks in the region.

Simon Feeny
September, 2013

Acknowledgements

This book has benefitted immensely from a large number of people from Australia, the Solomon Islands and Vanuatu. The book's contributors are very grateful for the feedback, advice and guidance they have received from numerous discussions, interviews, seminars and conference presentations.

In particular, we would like to thank the Australian Agency for International Development (AusAID) which generously supported our project and this book through their Australian Development Research Award (ADRA) scheme. We are also indebted to the teams of Solomon Islanders and Ni-Vanuatu for their tireless efforts in assisting us in conducting the research fieldwork. Special thanks must also go to the Oxfam staff and volunteers who have given up so much of their time (often for free) to make this book possible.

We are grateful for the input from AusAID staff including Gordon Burns, Lucy Carlson, Rob Christie, Stephen Deklin, David Green, Paul Greener, Patrick Haines, Rob Harden, Steve Hogg, Jane Lake, Jacqueline Lees, Scott McLennan, Anna McNicol, David Momcilovic, Inge Stokkel and Bernie Wyler.

Thank you very much to all Oxfam staff including Gibson Ala, Hannah Cropley, Kethy Cyrus, Neverlyn Efo, Alice Eric, Elizabeth Faerua, Alexandra Gartrell, Naomi Godden, Natalie Gray, Katie Greenwood, Alfred Kiva, Alex Mathieson, Chrisanta Muli, Thelma Namusu, Martha Mangale Rafe, Lorima Tuke, Jo Weber, Megan Williams and Nelly Willy.

Special thanks goes to the teams which helped us conduct the fieldwork in the Solomon Island including Paula Aruhuri, Moses Au, Johnny Avock, Wendy Beti, Elton Etega, Julia Fationo, John Filau, Shaniela Filiau, Charles Gane, Julia Garina'au, Jenny Inimae, Hilda Kii, Charles Konai, Ben Lesibana, Olson Luiramo, Esther Luvena, Captain Manuel of World Vision, James Mao, Teiba Mamu, David Manetiva, Laurie Nodua, Hexley Ona, Grace Orirana, Solomon Rakei Seimonana, Anna Grace Suhusia, Lenox Susukahu, Stivo Tabo, Apollonia Talo, Jessica Theokheranga, Thomas Tego, Sharon Tohaimae, John Toska, Sonta Usuli, Newman Wilie and Ckemeron Willie

Special thanks also goes to our teams in Vanuatu including Willy Albert, Harris Apos, Stynette Bakeo, Gibson B. Bani, Barry Boe, Mark Adin Boe, Alice Bolenga, Joe Bule, Annie Cyrus, Patricia Cyrus, Dalida from Hog Harbour, Iven Joshua, Alick Langitong, Nancy Matan, Sonia Mera, Johnny Nimau, Alick John Noel, Isabel Petri, Thomas Putunleta, Noe Saksak, Anita Samana, Dreli Solomon, Masten Tias, Anthea Toka, Simon Torle, Shamina Ulnaim, Annah Vera Vota, Daniella Woiala.

We are also grateful to a number of other people for their time, assistance and advice including Edith Bowles, Jared Berends, Derek Brien, Tim Bulman, Andrew Catford, the Economics Society of the Solomon Islands, Alison Fleming, Luke Forau, Nick Gagahe, Tobias Haque, Alphea Hou, Anthony Hughes, Erik Johnson, Simil Johnson, Michael Kikiolo, Benuel Lenge, Ruth Maetala, Sandra Moore, Chief Mor Mor, Anna Naemon, Benneth Ngwele, Don Patterson, Emma Peppler, Bob Pollard, Ralph Regenvanu, Rebekah Ramsay, Geoff Robinson, Peter Samuel, Patrick Shing, Jamie Tanguay, Michael Taurakota, Odo Tevi, Morrisen Timatua, Peter Toa, Paula Uluinaceva, John Usuramo, Andrew Waleng, Heather Wallace and Nancy Wells.

Finally, a very special thank you to a number of communities for their participation in the research as well as for their warm hospitality. In the Solomon Islands these communities included White River, Burns Creek, Lilisiana, Ambu, Uzaba, Pusisama, Maruiapa, Oa, Binu, Foxwood, and Malu'u. In Vanuatu the communities included Ohlen, Blacksands, Sarakata, Pepsi, Nerikniman, Todolak, Ra Island, Baravet, Hog Harbour, Mangalilu and Lelepa.

List of Abbreviations

ADB	Asian Development Bank
AusAID	Australian Agency for International Development
BMI	Body Mass Index
CBOs	Community Based Organisations
CCTs	Conditional Cash Transfers
EVI	Economic Vulnerability Index
FAO	Food and Agricultural Organisation
GDP	Gross Domestic Product
GEC	Global Economic Crisis
GFC	Global Financial Crisis
GNI	Gross National Income
GPPOL	Guadalcanal Plains Palm Oil Limited
HDI	Human Development Index
ILO	International Labour Organisation
LDCs	Least Developed Countries
MDGs	Millennium Development Goals
MICS	Multiple Indicator Cluster Survey
MMPI	Melanesian Multidimensional Poverty Index
MNCC	Malvatumauri National Council of Chiefs
MPI	Multidimensional Poverty Index
MSG	Melanesian Spearhead Group
NGOs	Non-Government Organisations
OPHI	Oxford Poverty and Human Development Initiative
PICs	Pacific Island Countries
PIFS	Pacific Islands Forum Secretariat
PNG	Papua New Guinea
PQLI	Physical Quality of Life Index
RSE	Recognised Seasonal Employer
SIDS	Small Island Developing States
SWP	Seasonal Workers Program
UNDP	United Nations Development Program
UNICEF	United Nations Children's Fund
VANWODS	Vanuatu Women's Development Scheme

Chapter 1
Household Vulnerability and Resilience to Economic Shocks in Melanesia: An Overview

Simon Feeny and May Miller-Dawkins

1.1 Introduction

Shocks are sudden events which impact on well-being. Covariate shocks are events that have impacts on entire countries or communities and include natural disasters as well as economic shocks such as price hikes and recessions. Idiosyncratic shocks affect only the household and include, for example, the death or illness of a family member or the loss of a job.

A single definition of vulnerability does not exist. It is a multifaceted, multidimensional concept. There is, however, general agreement that vulnerability is a forward-looking, or *ex ante*, measure of well-being. It therefore differs from the concept of poverty, which assesses current (rather than future) well-being status. At a household level, vulnerability is often defined as the likelihood or risk of being poor (however defined) or of falling into poverty in the future. Since different measures of poverty exist, vulnerability is conceptualized in different ways. In addition to being a function of the characteristics of risk or a shock and the ability of a household to respond to shocks, vulnerability also reflects relational considerations – between women and men, the young and old, people of different abilities and between different social or ethnic groups. Importantly, to the extent that that a households' vulnerability is also a function of its ability to recover once a shock has occurred it cannot be studied in isolation from resilience. The two concepts are inextricably linked. An extensive overview of the use of these terms is provided in Chapter 2.

This chapter focuses on the vulnerability and resilience of Melanesian and Pacific countries at a national level before presenting the findings of research conducted at the household level in two Melanesian countries: the Solomon Islands and Vanuatu. Other independent Melanesian countries include Papua New Guinea and Fiji. There are important differences among these countries, and some degree of caution must be exercised in applying the findings and implications from this book to the rest of Melanesia.

All Melanesian countries are Pacific Island Countries (PICs) that are often referred to as being among the most vulnerable in the world, in terms of both their exposure to risks and their ability to manage them.[1] In particular, they are susceptible to the impacts of natural disasters, climate change, commodity prices, and global economic downturns. A lack of research examining vulnerability at the household level in this region is therefore very surprising.

Data from EM-DAT (2012) indicate that PICs have experienced 105 natural disasters since 2000. Of these disasters, 70 occurred in Melanesian countries and affected over half a million people. The United Nations University's Institute for Environment and Human Security (UNU-EHS) finds that the four Melanesian countries are among the 20 countries in the world that have the highest probability of experiencing a disaster. In the Disaster Risk Index, Vanuatu is ranked first and the Solomon Islands fourth (UNU-EHS, 2011). The region is also one of the most vulnerable to climate change, which threatens agriculture, food security, health, the availability of fresh water and the existence of coastal settlements. The potential human and economic costs of climate change in the Pacific region are catastrophic.

Similar to countries across the world, PICs have recently experienced a triple crisis with price hikes for both food and fuel in 2007 and 2008, preceding a Global Economic Crisis (GEC) in 2009. During this period, households are likely to have felt the rise in the cost of living acutely. Among low income households in the Pacific, almost 50 per cent of household expenditure is on food, even in rural areas (Abbott, 2008). Moreover, most PICs are heavily dependent on imported fuel for energy and transport. Rises in fuel prices make transport more expensive and increase the cost of getting goods to market, particularly for the majority of the population that lives in rural areas and outer islands.

While the dominance of semi-subsistence livelihoods reduces vulnerability to food insecurity in Melanesian countries, McGregor et al. (2009) document acute vulnerability to food insecurity in low-income urban households in these countries. Moreover, the World Bank (2010) notes that, in the context of the Solomon Islands, the development bias towards the capital Honiara and to natural resource enclaves means that outer regions of the country are at a considerable disadvantage, since their remoteness implies high costs of providing infrastructure and services. The resulting lack of transport and communications networks severely constrains the movement of goods and people. In rural areas, household budgets are quickly squeezed if the price of food or fuel increases suddenly.

PICs were not immune to the impacts of the GEC. Despite large informal sectors, small manufacturing sectors and a stable banking system with limited exposure to global financial markets, PICs were vulnerable to the sudden downturn in demand owing to their undiversified economies Some are highly dependent upon tourism and capital inflows for foreign exchange and many have a dependence on food, fuel and capital goods (Chhibber et al., 2009). At a macroeconomic level, PICs were, to varying degrees, affected by the crisis through declining government

1 Independent Pacific Island Countries include the Cook Islands, Federated States of Micronesia, Fiji, Kiribati, Marshall Islands, Nauru, Niue, Palau, Papua New Guinea, Samoa, Solomon Islands, Tonga, Tuvalu and Vanuatu.

revenues from trade and other taxes, a loss in the value of trust funds (invested on international markets) and, in some cases, a fall in revenue from tourism. Many PICS are once again experiencing high rates of inflation and significant increases in food and fuel prices as these commodities recover from the GEC. At a regional level, while growth is expected to have picked up in 2011 (largely driven by the resource rich economies of Papua New Guinea and Timor-Leste), it is forecast to be lower in 2012 and 2013 (ADB, 2009a, 2009b, 2011, 2012, Feeny 2010).

There is very little information about how these recent shocks have affected individuals, households and communities in Melanesia. In the absence of formal monitoring systems, evaluating the impacts at this level remains very challenging. Extrapolating human impacts from macroeconomic changes often obscures their differential effects for women and men, and in the informal and reproductive economy (Green et al., 2010). These often weak linkages between the macro-economy and the household are compounded in PICs by the low level of participation in the formal economy and the reliance on subsistence and family support structures (Feeny 2010).[2]

Moreover, predicting any long-term impact on development, without information on how households respond when shocks occur, is virtually impossible. In particular, there is limited information, beyond stylized views, on the extent to which the custom (or *kastom* in the local languages of Pidgin and Bislama) economy acts as a crucial resilience mechanism in the event of shocks. Regenvanu (2009) considers the custom or traditional economy to be a key resilience mechanism that is unique to Melanesian countries, with a large proportion of households residing in rural areas, with good access to land on which to grow food and strong familial ties. Yet with rapid urbanization and monetization these safety nets could be deteriorating.[3] In response to these challenges, this chapter draws heavily on findings from research fieldwork that was specifically designed to provide insights into these issues.

2 There are a number of anthropological studies examining responses in the Pacific to natural disasters. For example, Firth (1959) examined lifestyle persistence in Tikopia, the Solomon Islands, by revisiting a community in 1952 that was originally visited in 1929. Firth documented the responses to a hurricane, drought and subsequent food shortages. Currey (1980) examines the impact of famines across the region noting that famine accounted for a large proportion of the migration of villages on the weather coast of Guadalcanal in the Solomon Islands. He also found that famines across the Pacific sometimes led to a change in the types of crops grown by villagers. Lieber (1977) provides a number of studies from the Homer Barnett project which studies relocated communities in the Pacific. These studies tend to stress a reliance on kinship more than the chapters within this collection, possibly reflecting shifts towards the individual with the increasing integration of communities into the formal market system.

3 While high rates of urbanisation have been a feature of most Melanesian countries, figures from the 2011 preliminary census of Papua New Guinea indicate that population growth for the capital Port Moresby has slowed considerably (see Bourke, 2012).

It reveals that in both rural and urban areas, households have a high degree of exposure to food and fuel price shocks as well as to environmental shocks such as natural disasters. Further, households have had to cope with the impacts of a number of idiosyncratic events. The food garden and the nearby coral reefs are clearly fundamental in providing resilience to households as is the support of kinship networks to cope with the effect of shocks.

The research finds important differences in the experience of shocks across location. A simple rural-urban disaggregation in the Solomon Islands and Vanuatu is not appropriate to explain how lifestyles are changing rapidly in these countries. Given their strong ties to the formal economy, urban households are highly exposed to price shocks and a lack of access to land threatens their food security and puts a strain on traditional resilience mechanisms. While it remains the case that many rural communities suffer from a lack of infrastructure, services and income earning opportunities, their access to land and subsistence agriculture often provides them with a buffer during price shocks. Yet there is strong evidence that rural communities with both good access to land and infrastructure that connects them with population centres are faring very well, with them benefiting from a combination of access to essential services, formal employment and international markets. This is discussed further in Chapter 5.

The remainder of this chapter is structured as follows. Section 1.2 provides further details on the fieldwork activities and the information collected from the Solomon Islands and Vanuatu. Section 1.3 broadly examines vulnerability by using the household survey data to summarize the different types of shocks that communities in Melanesia have experienced during the past few years. Section 1.4 addresses resilience by evaluating the impacts of these shocks and how households responded to them. Section 1.5 summarizes the various strategies that focus group participants recommended for reducing vulnerability and strengthening their resilience. Finally, Section 1.6 outlines the remaining themed chapters that comprise this book.

1.2 Research Fieldwork: Household and Community Data

This book comprises chapters that draw on unique data and information collected during extensive fieldwork in the Solomon Islands and Vanuatu. The fieldwork was conducted during 2010–11 and consisted of over 1,000 household surveys, more than 50 focus group discussions and a number of key informant interviews. Careful consideration was given to the location of fieldwork. In an attempt to capture the diversity in experiences of vulnerability and resilience, six locations were targeted in each country. These were selected based on criteria that sought to reflect diversities in remoteness, economic activity and the environment. The locations of the communities and their common characteristics are provided in Table 1.1.

Table 1.1 Research fieldwork locations and their characteristics

	Vanuatu	Solomon Islands	Characteristics
Urban	Port Vila (Efate) (Ohlen and Blacksands)	Honiara (Guadalcanal) (White River and Burns Creek)	Settlements in each country's capital city
	Luganville (Santo) (Pepsi and Sarakata)	Auki (Malaita) (Lilisiana and Ambu)	Settlements in each country's second largest town
Rural	Baravet (Pentecost)	Guadalcanal Plains Palm Oil Limited (GPPOL) Villages (Guadalcanal)	Rural communities heavily involved in commercial agriculture
	Hog Harbour (Santo)	Malu'u (Malaita)	Rural communities separated from the respective second city by a 100 km road
	Mangalilu/Lelepa Island (Efate)	Weather Coast/ Marau Sound (Guadalcanal)	Communities on the same island as the respective capital city with known links to Oxfam Australia.
	Mota Lava (Banks Islands)	Vella Lavella (Western Province)	Remote communities a significant distance from the respective capital cities

Source: Feeny et al. (2013).

In Honiara in the Solomon Islands, two communities, White River and Burns Creek, were visited. White River, known for its betel nut market, is a large and established settler community located in western Honiara. It has a broad mix of ethnicities, encompassing settlers from as far away as Temotu Province and Kiribati. Burns Creek is a squatter settlement in the east of Honiara, housing mostly displaced Malaitians from the recent ethnic conflict (1999–2003) known colloquially as 'the tensions'. This settlement has been characterized by high unemployment, gangs, and, sometimes, violent crime.

Auki is the provincial capital on the neighbouring island of Malaita and is the country's second largest town. The fishing villages of Lilisiana and Ambu, which mainly comprise villagers from Langa Langa Lagoon in the chain of artificial islands along the southern region of Malaita, were visited. Both communities are pressed up against the coast and experience flooding during king tides. Communities around Malu'u, the provincial substation in North Malaita about 80 km north from Auki, were also surveyed. A densely populated area, with hospital and schools, Malu'u is also renowned as a copra growing area. However, transport to the main markets in Auki is currently impeded by the very poor quality of the road.

The area around the Guadalcanal Plains Palm Oil Limited (GPPOL) factory is the major growing region for palm oil in the Solomon Islands. In addition to leasing their land or being employed on the major estates, members of some communities in the area have opted to become out-growers using production from small blocks of land to supply palm fruit to the GPPOL operations. These 176 blocks vary in size from less than 1 hectare to over 22 hectares (Fraenkel et al., 2010). In recent years, households have been able to make considerable amounts of money from out-growing.

The communities of Oa on the Weather Coast and Mariuapa in Marau Sound are located on the southeast corner of Guadalcanal. Communities in this part of the country are isolated, with very limited transport links to the capital (though the area does benefit from a small amount of tourism). Finally, small communities in Vella Lavella, located in the country's Western Province, were visited to examine the extent to which geographically distant communities in Melanesia are integrated into the global economy.

In Vanuatu, an attempt was made to select broadly comparable communities to the Solomon Islands. In the capital city, Port Vila, two settlements were visited: Ohlen and Blacksands. Both communities are home to migrants from the outer islands. Ohlen is a densely populated area, situated on an escarpment that also serves as Port Vila's water catchment area. It has few services and there are concerns that residents could face eviction. Blacksands is Port Vila's largest and most established informal settlement and has often been characterized by tensions, conflict and insecurity over land.

The two communities of Pepsi and Sarakata were visited in Luganville, Vanuatu's second largest town on Espiritu Santo. While only separated by the Sarakata river, these migrant communities have very different access to government services.[4]

The communities of Baravet, on the south west coast of Pentecost, were selected on the basis of their heavy reliance on cash crop farming, in particular kava, which is shipped to the main markets of Port Vila. Lebot et al. (1997) note that around two-thirds of kava shipped to the main markets of Port Vila and Luganville originated in Pentecost, and that farmers were rapidly transitioning

4 At the time of the survey, the Pepsi community was not considered to be under the auspices of the Luganville Municipal Council. Therefore, unlike Sarakata, it had no access to public utilities.

away from growing kava as a garden crop and into more industrialized growing processes, with dedicated growing areas.

Mangalilu (and the nearby island of Lelepa) are in north Efate – the same island as the capital. While rural in nature, these communities have good transport links to Port Vila following the recent completion of the island's ring road – a large US-backed infrastructure project undertaken by the Millennium Challenge Corporation (MCC). The selection of Hog Harbour reflected its similarity to Malu'u in terms of its proximity to the second city and its involvement in copra farming. However, unlike Malu'u, Hog Harbour is provisioned with good access to Luganville, following the completion of the MCC-sponsored East Santo Road, and access to tourism owing to its proximity to the renowned Champagne Beach. Similar to the Solomon Islands a geographically distant community was selected: that of Mota Lava in the Banks Islands in the far north of the Vanuatu archipelago.

In each location, a survey specifically designed to capture the experience of shocks and household responses to them was conducted for approximately 85 households. The survey design drew on an extensive literature survey of economic vulnerability and resilience, particularly in PICs. In addition to capturing information on households' income sources, asset endowments and demographic characteristics, it included questions on households' experiences of shocks and their responses to them. In the interests of parsimony the survey considered a 'household' to correspond to the dwelling. The survey deliberately avoided concentrating on the 'head' of the household because in pilots it was discovered that the head was repeatedly identified as the oldest male in the family – irrespective of whether that person resided in the dwelling itself – and due to evidence in the literature that 'head of household' surveys tended to exclude the views and experiences of women. One adult was therefore asked to answer questions on behalf of 'the people living in the house'. In an attempt to draw on the experiences and perspectives of both men and women, the research teams aimed for a gender balance in survey respondents, and 10 per cent of households were re-surveyed so that there were responses from both male and female respondents from within the same household.

In each site four separate focus group discussions, led by female or male local facilitators, were held: one each with men over the age of 30; women over the age of 30; younger men; and younger women. The focus group outline was more open than the survey – allowing groups to define what 'difficult' and 'good' times had been in the past and how they had responded. This provided a local definition of significant events and hardship against which the assumptions of the survey could be tested. Key informant interviews were held with government officials, aid agencies, community leaders, and village chiefs to further triangulate results.

Local research teams – team leaders, focus group facilitators and documenters (men and women), and surveyors – were recruited in each country and a gender balance was strictly adhered to within the teams. Teams participated in four days of training in the respective capital city with a strong focus on research ethics, methods and the survey and focus group outline. Teams were accompanied and assisted by a member of the research team and local Oxfam Australia staff.

To assist with the research fieldwork, 'point people' local to each research site were recruited in order to negotiate access to the communities and provide information to them prior to the arrival of the research team. On arrival in a research site, the teams would provide community briefings to inform them about the research, answer questions and start to arrange the best times for focus groups. Participants were provided with a copy of their consent form and contact details should they wanted to withdraw their consent. Prior to their departure, the teams held community de-briefings, outlining some of their initial findings and answering any further questions.

1.3 The Experience of Shocks in the Solomon Islands and Vanuatu

The household survey asked specific questions on each household's experience of covariate and idiosyncratic shocks. Responses to these questions are summarized in Figure 1.1. Virtually all households had experienced at least one kind of shock during the previous two years. Shocks can be categorized into: (i) economic shocks (which include covariate price shocks as well as reduced employment); (ii) natural disasters which usually impact on communities; and (iii) other shocks including crime and theft, death/illness of a household member and crop failure which are idiosyncratic in nature, with impacts on specific households.

Virtually all households had experienced at least one kind of shock during the previous two years. The most prevalent shocks relate to food and fuel price hikes. A household experienced a food/fuel price shock if, during the two years preceding the survey, they have experienced an increase in the price of food/fuel.[5] Almost 90 per cent of households experienced a food price shock and over 70 per cent a fuel price shock.

There is little distinct pattern to the experience of food price shocks across country and communities. Many rural communities appear just as likely to experience a food price shock as urban communities – reflecting the fact that most households purchase at least some of their food intake. Moreover, fuel price hikes will have increased the transport cost of getting food to rural locations and outer islands so rural communities may be paying more for the same food items. According to the ADB (2009c), fuel price rises can be magnified at each link of a

5 Households were presented with a Likert scale to measure whether, over the past two years, the price of food/fuel had: (i) gone down a lot; (ii) gone down; (iii) stayed the same; (iv) gone up; (v) gone up a lot. A food/fuel price shock could also be defined as whether a household has found, over the past two years, that the ease of obtaining food has become 'harder' or 'much harder' on a similar Likert scale. The figures for a household's experience of a food/fuel shock remain largely unchanged if this alternative definition is adopted.

very long value chain out to the most remote communities and are subsequently passed on to consumers – particularly for those foods that arrive by ship. This may partly explain why a high proportion of households in the geographically distant Banks Islands, experienced a food price shock – despite households having ample access to land for gardening. Food price shocks are also particularly prevalent in urban areas in both countries, as well as in Baravet in Vanuatu. Conversely, the GPPOL villages and households located in Hog Harbour had the lowest proportion of households experiencing a food price shock, possibly reflecting a good availability of land for gardens in these communities and them benefitting from out-growing and tourism. A food price shock was also less commonly experienced by households on the Weather Coast, which is likely to reflect these communities' genuine remoteness, and thus relative reliance on food from the garden rather than the store.

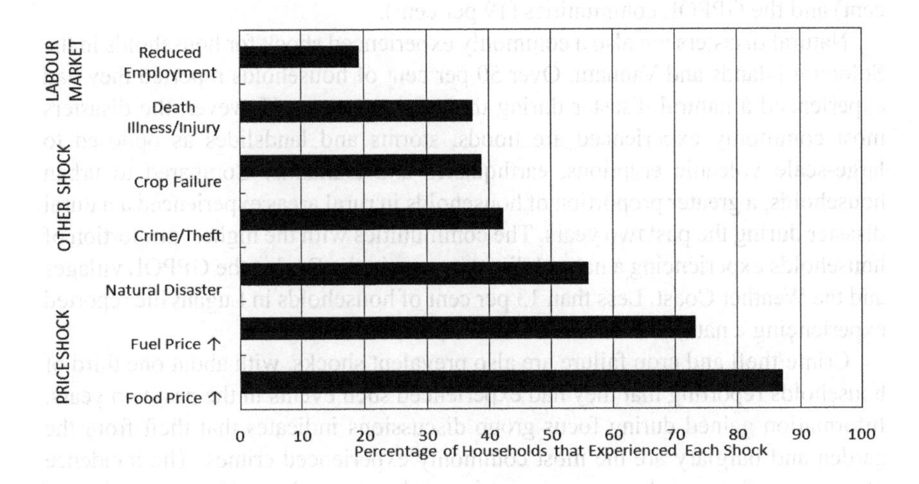

Figure 1.1 Shocks most commonly experienced by households in the Solomon Islands and Vanuatu

Note: Sample size (N) = 955 households

Source: The authors.

In the Solomon Islands, the highest proportion of households experiencing a fuel price hike was found in Honiara and the isolated communities of the Weather Coast.[6] In Vanuatu, the same was true for households located in Luganville and

6 Although a large proportion of households on the Weather Coast reported that fuel prices had 'gone up' only a small proportion reported that they had 'gone up a lot'.

the Banks. These findings reflect the high use of fuel in both urban and remote areas of Melanesian countries. Fuel is required extensively for public transport, in particular buses, in urban areas but there is also the need for fuel in rural areas for energy, cooking and transport in order to access health and education services. In rural areas, fuel is often purchased at a much higher price. The Solomon Islands communities of Malu'u and Vella Lavella had the lowest proportion of households reporting a fuel price shock.

Interestingly, reduced employment, which is how a macroeconomic shock is most often transmitted to households, has not been commonly experienced by households during the past two years. Reduced employment is defined as a household member losing their job or having their hours reduced. About 18 per cent of households report experiencing this type of shock. The highest proportion of households reporting this shock can be found in areas with the highest rates of formal sector employment: Port Vila (28 per cent), followed by Honiara (24 per cent) and the GPPOL communities (19 per cent).

Natural disasters are also a commonly experienced shock for households in the Solomon Islands and Vanuatu. Over 50 per cent of households reported they had experienced a natural disaster during the past two years. However, the disasters most commonly experienced are floods, storms and landslides as opposed to large-scale volcanic eruptions, earthquakes and tsunamis. Compared to urban households, a greater proportion of households in rural areas experienced a natural disaster during the past two years. The communities with the highest proportion of households experiencing a natural disaster were in the Banks, the GPPOL villages and the Weather Coast. Less than 13 per cent of households in Luganville reported experiencing a natural disaster.

Crime/theft and crop failure are also prevalent shocks, with about one third of households reporting that they had experienced such events in the past two years. Information gained during focus group discussions indicates that theft from the garden and burglary are the most commonly experienced crimes. The incidence of crime varies greatly across countries and communities. The proportion of households reporting the theft of livestock or crops in Vanuatu was a startling 48 per cent, compared to just 22 per cent in the Solomon Islands. More than 70 per cent of households in the Banks reported theft, with high rates also reported by households in Port Vila, Mangalilu and Hog Harbour. Communities in Auki and the Weather Coast were virtually crime free with less than 5 per cent of households reporting theft. Crop failure is another shock commonly experienced by households in the Solomon Islands and Vanuatu with nearly 40 per cent of households experiencing crop failure in the past two years. It is most common in rural areas given the greater prevalence of the garden and land for cash crops.

In general, responses to experiencing different shocks did not vary greatly across male and female respondents. However, exceptions are that women were more likely to report food price hikes and they also experienced greater difficulty in being able to manage the rise in food prices. As a consequence, women more often reported reduced consumption of food, concern about family health and nutrition, and increased reliance on the garden as food prices rise. This is likely to be reflective of women's usual responsibility for purchasing the households' food. This is discussed further in Chapter 3.

It is important to note that households can experience positive as well as negative shocks. Findings from the household survey reveal that 27 per cent of households experienced a positive economic shock, whereby the household received more money than they expected, primarily from obtaining more work, getting paid more, growing additional crops or receiving remittances. Households most commonly use windfalls to save, spend money on education or improve the quality of their house.

Examples of events that might lead to households experiencing a positive economic shock include donor funded roads in east Santo and around Efate in Vanuatu and (global) price hikes in copra and palm oil. Roads and improved transport infrastructure allow for the cheaper transportation of goods (and people) to market. In some areas households are responding to higher commodity prices by increasingly engaging in cash cropping and commercial agriculture. While this provides a higher income, it comes at the expense of spending less time maintaining their garden. Households and focus group participants often reported neglected gardens being destroyed by wild pigs.

1.4 The Impacts and Responses to Shocks

The impacts and responses to shock are inextricably linked. Information on how shocks have affected peoples' lives is best captured by the focus groups participants, while gauging the relative frequency of different responses is best achieved through household surveys. On average, households suffered five shocks in the preceding two years.

In general, focus group participants reported that the shocks they had experienced had made it harder to make ends meet. Households experienced an inability to generate a sufficient income in order to maintain an adequate standard of living and this restricted their ability to access markets, health clinics and schools. Health and education facilities can be a long distance from where households live, a problem that is exacerbated by poor infrastructure. The stress of having inadequate income is leading to broader social impacts including crime, violence, substance abuse and mental health problems. In focus groups, women consistently raised significant difficulties with access to sanitation and their experience of violence; issues that were not raised or emphasized in the same way by men.

1.4.1 Who Are the Most Vulnerable to Shocks and How Do They Experience the Impacts?

Focus group participants helped identify specific groups which are most vulnerable to the impacts of shocks. In general, urban communities were cited as being more vulnerable to price shocks, given their relatively greater reliance on imported foodstuffs than rural communities, which compounded the burden on households to find the money needed to pay bills for electricity and water as well as for transport. The greater access that people have to land in rural areas is able to provide them with a buffer when prices increase.

However, results from the household survey nuance this perception – as many rural communities were also exposed to rising food prices. In both rural and urban alike, people experienced higher prices for fuel: predominantly through higher transport costs. In urban areas there is a heavy dependence on public buses while trucks and boats are more commonly used in more remote locations. This makes it harder for people to transport goods, access jobs and markets; and it is more expensive to fish using outboard motors.

For households in urban areas, a lack of access to land for gardening clearly impacts on their vulnerability and resilience. High rates of urbanization and population growth are adding to the pressure on land, both in terms of tightening supply and through environmental degradation. Where land is available its ownership is sometimes disputed. Where plots of land do exist for gardening in urban areas they are often declining in size due to migration to urban areas. Households in urban areas are growing in size with family members moving to urban areas and increasing the reliance on a garden as a source of food. The result is that urban households are increasingly finding the garden to be inadequate as a food source and, as a consequence, are sometimes moving towards lower cost, imported food (such as rice, tinned fish and packet noodles). While there is a danger that such moves will have negative health implications, it is not necessarily true that processed and imported foods are of lower nutritional value. Assertions regarding nutrition must take into account the broader issue of current diets and the dietary requirements of households.[7] That said, alarmingly, sometimes families are going hungry. In contrast, while rural households will nearly always get a certain proportion of their food from the local store, they generally have relatively good access to a productive garden, which increases their resilience to food price hikes given that there is a readily available substitute for sourcing their nutritional needs. However, rural households are still acutely affected by increasing costs and they remain constrained by a lack of income-earning opportunities.

Women have been bearing a substantial amount of the costs of adjustment as a result of these shocks (discussed in detail in Chapter 3). As a response, the workload of women has increased as they have borne the responsibility of finding additional income – while maintaining their traditional gendered responsibilities around the home. Women indicated that in order to make ends meet they had

7 For example, Allen (2001) finds that food imports improved food security in Malo, Vanuatu. In the context of Papua New Guinea (PNG), Heywood and Hide (1994) found that moves to cash cropping led to higher incomes and more spending on rice, tinned fish and meat. This resulted in beneficial increases in the intake of protein among households. Bourke and Harwood (2009) provide an extensive overview of food and agriculture in PNG.

increased their production from the garden, made handicrafts and cooked additional food to sell at local markets. Further, women are susceptible to domestic violence and some focus groups in urban areas even reported that women can be forced into prostitution. Although women focus group participants discussed increased feelings of insecurity at home as well as vulnerability to violence and rape, it cannot be determined that this is a direct result of economic shocks; rather, women's focus groups raised the issue of violence against women and advocated the need to address this persistent issue urgently during discussions about their difficulty in coping with shocks.

Youth was identified by focus group participants as another vulnerable group as young people have high rates of unemployment and a lack of income-earning opportunities. This was substantiated by the household survey, which indicated that a smaller proportion of men under the age of 30 were employed relative to older men. Sometimes, when young men successfully find income-earning opportunities, this challenges the traditional dominance of older men in the household and leads to tension. Squatter communities on insecure land are another vulnerable group. With insecurity over their land such communities are often reluctant to establish gardens, significantly reducing their resilience when shocks do occur. Widows, old people and those with a disability are also vulnerable since they are likely to rely greatly on other people to meet their needs.

Focus group participants identified the least vulnerable households as those with working members that can provide other household members with remittances. Along these lines, households with members that participate in Australia's Seasonal Workers Program or New Zealand's Recognised Seasonal Employer (RSE) scheme are perceived to be less vulnerable. Households with well-educated members are also perceived to be less vulnerable since they stand a better chance of finding employment. Further, most communities, particularly urban communities, are experiencing rapid population growth and smaller families with fewer mouths to feed were cited as being less vulnerable and more able to cope in times of hardship.

1.4.2 How do Households Respond to Shocks?

In collecting household responses to the different types of shock experienced, the survey found that households respond in similar ways to the different events. For example, households that experienced a food or fuel price shock responded in the same way as when they experienced a natural disaster. Further, households were found to undertake multiple actions following a shock with an average of 11 different responses.

The most common responses to a food or fuel price hike are provided by Figure 1.2, which clearly highlights the importance of the garden as a source of resilience. Over 80 per cent of households turned to their garden in order to source more food. Not surprisingly, this was a less common response in the urban communities of Honiara, Auki and Luganville but a universal response in the rural communities of Baravet and Malu'u. It was common in Port Vila, reflecting the access that communities in Ohlen and Blacksands have to urban gardens, despite the insecurity of land tenure.

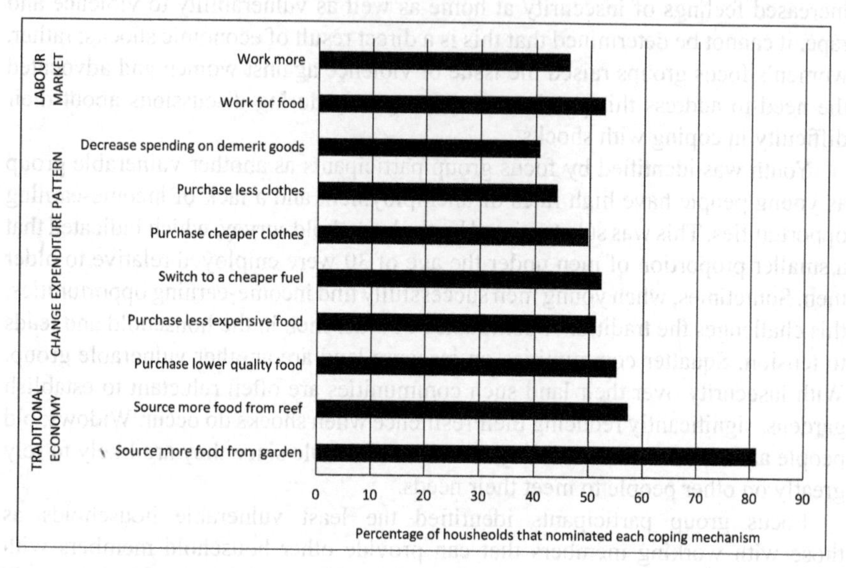

Figure 1.2 Most common household responses to a food or fuel price shock in the Solomon Islands and Vanuatu

Note: Sample size (N) = 872 households

Source: The authors.

Another important household response to a food or fuel price shock was to source more food from the reef, through various forms of fishing. While none of the sampled communities are far from the sea, this response was most common in the rural communities of Baravet and Vella Lavella and undertaken to a far less extent by households in Honiara and the GPPOL villages. We classify these garden and reef responses as traditional economy responses.

Overall, over 50 per cent of households in the sample responded to a food or fuel price shock by purchasing cheaper or lower quality food. This practice is more common in urban areas and was notably far less prevalent on the Weather Coast, reflecting this community being more reliant on food grown from the garden. More positively, approximately 42 per cent of households responded by reducing their consumption of 'demerit goods' which are unhealthy goods such as kava, betel nut, cigarettes and alcohol. Not surprisingly, this response was more widely undertaken in urban communities, in particular, Honiara, Auki and Port Vila with more widespread availability and use in these areas. However, 60 per cent of households in the rural location of Vella Lavella reported responding in this way.

Switching to a cheaper fuel was a much more common response in the Solomon Islands than in Vanuatu. Almost 60 per cent of households responded in this way in the former compared to less than 45 per cent in the latter. It was a particularly common response in GPPOL villages and the Weather Coast, possibly due to the wide availability of firewood in these communities. While responses such as fishing more and turning to firewood for fuel can provide resilience in the short term, reductions in these resources can actually undermine resilience of communities in the longer term.

Purchasing lower quality and fewer clothes was particularly common in Port Vila and Mangalilu, where people in these communities might purchase clothing more often than those located in rural and more remote regions. Working for food was a common response on the Weather Coast and the GPPOL villages. Clearly, there are good opportunities for work around GPPOL which neighbouring communities take advantage of when necessary.

From a long-term development perspective it is reassuring that 'no longer seeking medical help' and 'removing children from school' were not particularly common responses to a food or fuel price shock. Overall, less than 7 per cent of households did not seek medical help for an illness or injury following a shock, although this was more common in Vanuatu (10 per cent of households) than in the Solomon Islands (3 per cent of households). While this can be viewed in a positive light, it is also likely to reflect the already poor access that these communities have to health services. However, approximately 22 per cent of households reported buying less medicine, with a higher proportion of households responding in this way in the capital cities given the better access they have to modern pharmacies.

Primary school education is now officially fee-free in both the Solomon Islands and Vanuatu although families still have to meet the costs of materials, uniforms and transport and schools sometimes ask for their own fees. Only 9 per cent of households took their children out of school in response to shocks (this was the same for both the Solomon Islands and Vanuatu). Moreover, just 3 per cent of households sent a child out to work. More commonly, households are found to move their child to a cheaper schooling option with 29 per cent of households responding in this way. This move might involve sending a child to a closer school or to live with extended family members to reduce the cost of getting to school. This response was particularly common in the Solomon Islands.

The survey also found that 11 per cent of households reduced their savings, but far fewer households borrowed money (about 2 per cent) given the limited availability of financial services in these countries.[8] In response to a food or fuel

8 It is not uncommon for people in Port Vila settlements to offer informal 'payday loans' to one another. Further, finance is sometimes available through NGOs such as the Vanuatu Women's Development Scheme (VANWODS).

price shock about 18 per cent of households sold livestock or other assets to get food which could, unfortunately, make them more vulnerable to future shocks.

Households also responded by reducing their financial commitments to social networks (discussed further in Chapter 6). Around 19 per cent of household indicated that they provided less money to community fundraising events. Additionally, about 11 per cent of households said they gave less money to *wantoks* (*wantok* is a Pidgin word from Papua New Guinea and refers to people that speak the same language, are from the same place, or close friends and family). At the same time, households reported increasingly drawing on their social networks; sourcing help from friends (28 per cent), family (17 per cent) and neighbours (5 per cent).

1.5 What Policies and Strategies Can Best Assist Households Cope with Shocks?

As part of the focus group discussion exercises, participants were asked what they feel they can do, and what others can do, to help make things better for their community through a 'low hanging fruit' exercise. Low hanging fruit is the easiest fruit to pick from the tree and links with the idea that householders can easily do some things themselves. Fruit that is hanging higher on the tree is harder to get and obtaining it might involve help from others either in the community or from the government. Focus group participants identified four broad strategies that could reduce their vulnerability and increase their resilience: reducing reliance on imported foods; improving the ability to generate an income; improving education; and reducing social tensions and fragmentation. Each is discussed in turn and this section also draws on information from key informant interviews. Interestingly, while the traditional way of life in Melanesia provides resilience for communities, some of the strategies identified will actually embed them more in the formal economic system and increase vulnerability to future economic shocks.

Reducing their reliance on imported foods was common theme across all focus groups and can be achieved in a number of ways. Some focus group participants suggested that households should grow more food as well as fish more to improve their food security, despite recognition that this is increasingly hard for urban households. At a community level, participants identified a need for community cooperatives to assist with both the production and selling of locally grown food and to increase household incomes. Participants also identified a need for households, community leaders and the government to change attitudes towards food, promoting the production and consumption of traditional food. The need to change the attitude of children in particular was perceived to be very important. Rural focus group participants in the Solomon Islands wanted greater support from the Ministry of Agriculture at times of crop failure or natural disaster, such as flooding and participants of one focus group in Vanuatu believed that the government should subsidize the mass production of local food.

Focus group participants identified the ability to generate a sufficient income as fundamental to the capacity of a household to cope with price shocks. Here, participants asserted that government could assist in a number of ways, such as by improving job opportunities and raising the minimum wage. For example, in rural areas households reported reluctance to plant cash crops because of the volatility in their price. Participants felt the government could play a role by implementing price controls to provide them with greater certainty, and therefore more incentive, to produce. A minimum wage would entice more people into the formal sector. Government could also play a role in capacity building, providing information on prices, livelihood training and providing greater access to financial services to assist households to save. In general, it was felt that governments should assist with the establishment of small enterprises (rather than impede them through, for example, the banning of betel nut vendors in Honiara).

In more affluent communities, particularly those benefiting from cash cropping, households earn enough to meet the basic needs of the household and school fees and then stop working. While this is sometimes attributed, in part, to attitudes to *wantok* obligations, an alternative explanation is that households have nothing they can do with their excess money. Financial facilities could assist with saving, as long as fees are kept at reasonable rates.

The local community and households themselves can play an important role in income generation. Again establishing community cooperatives to pool resources, share information and reduce the costs of getting goods to markets that are further away could play an important role in income generation. Yet it should not be overlooked that, in the focus group discussions, a lot of emphasis was also placed on the individual or household for dealing with shocks, with little tolerance and acceptance of idleness within the community. 'Plant more, grow more and sell more' was a common mantra. Participants believed that household budgeting is currently inadequate to cope with shocks and could be improved. Again, this places the responsibility to improve things on the household. A related issue is for households to reduce their spending on non-essential items such as mobile phones, kava, alcohol and cigarettes.

Improved access to markets was also central in many focus group discussions. This is particularly important for women, as they are expected to engage in market activity and need to 'make more market' at certain times. High fuel prices and poor transport infrastructure makes travelling to markets very expensive in terms of both time and money. This is true for communities in both urban and rural areas. High transport costs are exacerbated by high market fees. Key informants spoke of a need for government to establish new marketplaces near key communities and for communities to pool resources to reduce the costs of getting goods to market. For instance, there is only one main market in both Honiara and Port Vila (although there are some, much smaller, roadside markets scattered across these towns). If there were additional markets on the edge of these cities improving the accessibility of markets would reduce both the need for and the cost of transport. Governments could also improve transport services. Better roads would enable communities to get produce to market a few times a week rather than facing the occasional very long walk.

The inaccessibility of education emerged as a common theme with focus group participants. They saw an education as a vital component of resilience by recognizing the links between a lack of education, unemployment, idleness, social tensions and crime. Despite being notionally free, the costs of sending children to primary school are proving very difficult for many households to meet. Since children often take public transport to school, fuel price hikes can have a direct impact on school costs, combined with those for stationery, uniforms and lunches. Participants identified an increased role for government in assisting households by providing free secondary schooling, providing boarding at schools in isolated areas and assisting with other costs associated with gaining an education. Again, though, households themselves can play a role by increasing their income and ensuring the children attend school. Improving access to, and the quality of, health services, water and sanitation will also assist in raising school attendance, particularly for girls. Some participants and key informants viewed population growth as a key factor in vulnerability and recommended education on family planning.

Finally, participants identified a need for government, communities and households to address rising social tensions, fragmentation, theft and substance abuse. Women – young and old – particularly raised issues of domestic violence as well as transactional sex and prostitution as young women are pressured to earn more money. Solutions emphasized improved law enforcement and punishment through official channels but also using traditional mechanisms, such as village courts and chieftain dictates.

Kava use among youth in particular is an issue requiring special attention. In addition to improved law enforcement, government support for more sporting facilities, youth programmes and rural training centres were spoken about as ways of getting youth to contribute to their communities and society.

1.6 Overview of the Book

This book comprises thematic chapters to provide a comprehensive assessment of the experiences of shocks and responses to them in the Solomon Islands and Vanuatu. These themes include meanings of vulnerability and resilience, gender and youth dimensions of shocks, migration and urban-rural differences in vulnerability and resilience, the meaning and measurement of poverty in Melanesian societies, and the role of the traditional, or custom, economy in providing resilience. The concluding chapter uses the book's findings to assess the case for formal social protection policies to be implemented in Melanesian countries.

In Chapter 2, McDonald explores the different definitions and interpretations of the terms 'vulnerability' and 'resilience' as well as the different approaches to assessing them. There is a consensus that vulnerability differs from poverty by being a forward-looking measure of well-being assessing the likelihood of experiencing poverty in the future. There is also a consensus that vulnerability and resilience are concepts which cannot be separated, with the latter being a key determinant of the former. At the same time, different disciplines often view people and households as being vulnerable to different situations. These include vulnerability to falling into (i) income poverty; (ii) below a critical threshold of food security; (iii) livelihood stress; (iv) circumstances of weaker social capital or relations; and (v) experiencing a natural disaster. Throughout his chapter, McDonald uses the metaphor of a coconut tree in a storm to simplify the often complex terminology surrounding vulnerability and resilience. This metaphor was also used in explaining the research to community members in research sites.

In Chapter 3, Donahue, Eccles and Miller-Dawkins examine the specific experiences of women to shocks in the Solomon Islands and Vanuatu. Analyses of economic shocks often miss gendered and age-related impacts due to an absence of relevant data and by omitting examinations of the informal and reproductive economies. These impacts are particularly important in examining shocks in Melanesian countries given the significant prevailing gender inequalities. This chapter highlights how the multiple roles of Melanesian women imply a great poverty of time. Women bear disproportionate responsibility for maintaining the household, caring for children, the sick and elderly, and sometimes fishing and tending to gardens. Moreover, gender violence and political participation in Melanesia are noted as being amongst the worst in the world. The chapter finds that the gendered responses to shocks have contributed to reinforcing existing gender inequalities that exist in Melanesian societies.

In Chapter 4, Posso and Clarke examine migration patterns in Melanesia, together with the related issues of urbanization and remittances. Melanesian people have not had the access to other nation's labour markets that some other Pacific countries have enjoyed and most migration is therefore internal. While there is a large literature on the impact of remittance from migrants overseas, the role of internal remittances (of cash, food and clothing) is under researched. Interestingly, the chapter finds a high level of migration from urban to rural areas. While this does not contradict the high rates of urbanization in these countries, it demonstrates the prevalence of 'circular' migration, with movements to and from both urban and rural areas. Mobility in Melanesia is therefore a better descriptor than migration to understand the patterns of human movement within this part of the world. By far the most prominent motive for migration is family reasons, highlighting the strength of family support in Melanesian life. Additionally, the authors find that significant percentages of urban dwellers rely on their *wantok* for remittances of food and clothing as a coping mechanism against hikes in the prices of fuel and food, as well as other economic shocks.

In Chapter 5, Feeny, McDonald and Clarke argue that the 'subsistence affluence' that has traditionally characterized Melanesian lifestyle is increasingly threatened and more Melanesians are becoming vulnerable to poverty. The chapter calculates the Multidimensional Poverty Index (MPI) for communities in the Solomon Islands and Vanuatu. The MPI (at a country level) is now reported in the

United Nations Development Program's annual Human Development Reports but has not been available for the Solomon Islands due to a lack of data. The MPI is a composite index of well-being including information on health, education and living standards. It measures the extent and depth of a number of deprivations at a household level. The chapter finds a large variation in the MPI across communities. The incidence of poverty is highest in urban and remote areas but is low in rural communities with good access to infrastructure and markets. Moreover, a large proportion of the population in these countries is close to the poverty threshold and therefore vulnerable to experiencing poverty. The chapter proceeds by tailoring the index, including an additional 'access' dimension in order to be more appropriate to the Melanesian context.

In Chapter 6, McDonald, Naidu and Mohanty explore the interactions between vulnerability, resilience and the traditional (or custom) economy in Melanesia. The custom economy often refers to people living a subsistence or semi-subsistence lifestyle on communally owned land. Customary economy is widely practiced in Melanesia providing resilience to households and communities in the face of global and household level shocks. Increasing monetization and interaction with the formal economy are, however, having an impact. Customary obligations can become an additional burden on a household as caring for, and financially supporting, extended family members proves difficult. Support is increasingly needed in the form of cash including at custom ceremonies. Meeting customary obligations can actually increase a household's vulnerability and can sometimes lead to destructive behaviour among household members.

In Chapter 7, Feeny concludes by summarizing the findings of the book and using these findings to make a case for formal social protection policies to be implemented in the Solomon Islands and Vanuatu. Many developing countries seek to protect the poor and vulnerable members of their societies through *inter alia* Conditional Cash Transfers (CCTs), child grants, pensions and disability benefits. While the case for implementing these policies in the Solomon Islands and Vanuatu is strong, there are a number of issues which need to be considered by policymakers in their design. A number of areas for future research are also identified by this chapter.

References

Abbott, D. (2008), *A Macroeconomic Assessment of Poverty and Hardship in the Pacific: Lessons, Challenges and Policy Responses for Achieving the MDGs*, United Nations Development Program (UNDP) Pacific Centre, Fiji.

ADB (2009a), *Pacific Economic Monitor: May 2009* (Asian Development Bank: Manila).

ADB (2009b), *Pacific Economic Monitor: November 2009* (Asian Development Bank: Manila).

ADB (2009c), *Taking Control of Oil: Managing Dependence on Petroleum Fuels in the Pacific* (Asian Development Bank: Manila).

ADB (2011), *Pacific Economic Monitor: July 2011* (Asian Development Bank: Manila).

ADB (2012), *Pacific Economic Monitor: March 2012* (Asian Development Bank: Manila).

Allen, M. (2001), *Change and Continuity: Land Use and Agriculture on Malo Island, Vanuatu* (MSc Thesis, Department of Geography, Australian National University: Canberra).

Bourke, M. (2012), PNG's Population Predictions Compared with Preliminary 2011 Census Data (Outrigger: Blog of the Pacific Institute, Australian National University) [accessed March 2013].

Bourke, R.M. and Harwood, T. (eds) (2009), *Food and Agriculture in Papua New Guinea*, (ANU E Press, Australian National University:Canberra).

Chhibber, A., Ghosh, J. and Palanivel, T. (2009), *The Global Financial Crisis and the Asia-Pacific Region – A Synthesis Study Incorporating Evidence from Country Case Studies* (United Nations Development Program (UNDP) Pacific Centre: Fiji)

Currey, B. (1980), Famine in the Pacific: Losing the Chances for Change, *GeoJournal*, 4(5): 447–66.

EM-DAT (2012), *Emergency Events Database*, Centre for Research on the Epidemiology of Disasters (CRED) (Leuven: Belgium)

Feeny, S. (2010), *The Impacts of the Global Economic Crisis on the Pacific Region* (Oxfam Australia: Melbourne).

Feeny, S., McDonald, L., Miller-Dawkins, M., Donahue, J. and Posso, A. (2013), Household Vulnerability and Resilience to Shocks: Findings from Solomon Islands and Vanuatu, SSGM Discussion Paper 2013/3. Canberra: State, Society and Governance in Melanesia Project, RSPAS, Australian National University.

Firth, R. (1959), *Social Change in Tikopia: A Re-Study of a Polynesian Community after a Generation* (Allen & Unwin: London).

Fraenkel, J., Allen, M. and Brock, H. (2010), The Resumption of Palm-Oil Production on Guadalcanal's Northern Plains, *Pacific Economic Bulletin*, 25(1): 64–75.

Green, D., King, R. and Miller-Dawkins, M. (2010), The Global Economic Crisis and Developing Countries, Oxfam Research Report (Oxfam International: UK).

Heywood, P.F. and Hide, R.L. (1994), Nutritional Effects of Export-Crop Production in Papua New Guinea: A Review of the Evidence, *Food and Nutrition Bulletin* 15(3): 233–49.

Lebot, V., Merlin, M., and Lindstrom, L. (1997), *Kava, the Pacific Elixir: The Definitive Guide to its Ethnobotany, History and Chemistry* (Healing Arts Press: Rochester, Vermont).

Lieber, M. (ed.) (1977), *Exiles and Migrants in Ocean*ia (University Press of Hawaii: Honolulu).

McGregor, A., Bourke, R.M., Manley, M., Tubuna, S and Deo, R. (2009), Pacific Islands Food Security: Situation Challenges and Opportunities, *Pacific Economic Bulletin* 24 (2): 24–42.
Regenvanu, R. (2009), *The Traditional Economy as the Source of Resilience in Melanesia* (Vanuatu Cultural Centre: Port Vila).
UNU-EHS (2011), *World Risk Report 2011* (United Nations University Institute for Environment and Human Security: Bonn).
World Bank (2010), Solomon Islands Growth Prospects: Constraints and Policy Priorities. Discussion Note, October 2010 (World Bank: Honiara).

Chapter 2
Coconut Trees in a Cyclone: Vulnerability and Resilience in a Melanesian Context

Lachlan McDonald

2.1 Introduction

> Two coconut trees stand tall on an island beach as a cyclone builds offshore. In time, the power of the cyclone will hit both trees equally. However, while one tree will succumb to the storm and fall, the other tree will withstand the effects of the cyclone and remain standing. This poses several important questions: Can one predict which tree is more vulnerable to falling? More broadly, what are the key factors at play that allow one tree to withstand the effects of a cyclone and cause the other to fall? Also, what steps can be taken prior to the cyclone making landfall to ensure that each tree has the best chance of withstanding the event?

The above metaphor encapsulates the predicament facing some of the most vulnerable people in Melanesia today. One could easily reinterpret these 'trees' as households, the 'cyclone' as an unavoidable adverse event (or shock) such as a significant rise in imported food prices and 'falling over' as falling below a socially acceptable minimum standard of living (that is poverty). While the tree that is likely to succumb to the storm is considered to be 'vulnerable' to the cyclone's effects (and the greater the chance that the tree will fall, the more vulnerable that tree is), so too are individuals or households vulnerable to the effects of unanticipated shocks that would see them fall into poverty or, if they are already poor, plunge deeper into poverty. In contrast, just like the tree that remains standing following the cyclone, those households that are able to withstand the effects of an adverse shock without falling into poverty are resilient.

Pinning down what vulnerability means in any context, let alone in Melanesia, is not straightforward. Chambers wrote that 'vulnerable and vulnerability are common terms in the lexicon of development, but their use is often vague' (Chambers, 2006, p. 33). According to Hoddinott and Quisumbing this reflects the fact that, 'vulnerability – like risk and love – means different things to different people' (Hoddinott and Quisumbing, 2003, p. 1). It is also likely to reflect the sheer preponderance of definitions and methodologies there are to characterize the concept of vulnerability. Vulnerability assessments are a feature of a number of different disciplines, including, *inter alia*, economics, food security, sustainable livelihoods, sociology and disaster management. This chapter attempts to summarize the various definitions and methodologies that these approaches take. It takes a somewhat different approach to other literature reviews (such as Alwang et al., 2001; Heitzmann et al., 2002; Hoddinott and Quisumbing, 2003), as it uses the metaphor of coconut trees in a storm to illustrate the multitude of dynamics at play. The metaphor was initially developed as a research tool and used in focus-group discussions as part of the broader project underpinning this book. As was the case in the focus-group discussions, it is hoped that the articulation of vulnerability through the use of an intuitive metaphor will provide readers with a new perspective on the various ways that different disciplines conceptualize and measure different dimensions of the same problem.

While most definitions of vulnerability tend to stay true to the Latin etymological root of the word vulnerability, 'vulnerare', or 'to wound', there are varying perspectives on what such a 'wound' represents and how it can be assessed. To be sure, there are a number of universal dimensions to vulnerability. Naudé et al. (2009) outline four: (i) vulnerability is a forward-looking concept and therefore has some predictive function; (ii) the measurement of vulnerability is generally compared to a benchmark minimum level of welfare; (iii) vulnerability generally relates to some particular hazard; and (iv) vulnerability cannot be viewed in isolation from resilience – that is, how effectively risk can be absorbed and coped with.

However, a settled definition of vulnerability, and an agreed-upon estimation method, has thus far remained elusive. Sumner and Mallett (2011) consider the definitional imprecision as both a weakness and strength of vulnerability as an analytic tool: while intellectual fragmentation of vulnerability has given rise to a multiplicity of interpretations, the very fact that different disciplines focus their attention on the same central concept is testament to its importance. The upshot is that reducing vulnerability into a single sentence or estimation method may not be warranted. Rather, as this chapter will suggest, it may be preferable for comprehensive vulnerability analyses to take a multidimensional perspective. The remainder of this chapter is structured as follows. Section 2.2 explains, using the metaphor, the external and internal dimensions of vulnerability and resilience. Section 2.3 highlights the multidimensionality of both vulnerability and resilience, by stepping through the most salient characteristics of each of the most prominent approaches to vulnerability, including economics, food security, sustainable livelihoods analyses, sociology and disaster management. Section 2.4 examines the multidimensionality of vulnerability in a Melanesian context and Section 2.5 concludes.

2.2 The External and Internal Dimensions of Vulnerability and Resilience

Vulnerability is the result of a complex interaction between risk, responses and outcomes. An object (be it a tree in a cyclone, a household facing an economic shock or an individual fighting illness) is said to be vulnerable *to* suffering an undesirable outcome and this vulnerability comes *from* exposure to risk (Alwang et al., 2001).[1] In the example provided above, each tree is vulnerable *to* the negative outcome of falling over, which results *from* their exposure to risk, which includes the damaging effects of the cyclone.

However, whether a tree falls in a cyclone is not solely dependent on the magnitude of the storm, it also depends upon the tree's capability to withstand the cyclone's effects. This illustrates the key concept of resilience. Adger formally defines resilience as 'the magnitude of the disturbance that can be withstood before a system changes to a radically different state' (Adger, 2006, p. 268). While the concept of resilience has its roots in the natural sciences (in particular ecology), both resilience and vulnerability are increasingly becoming part of the social sciences (Miller et al. 2010). That vulnerability and resilience are intertwined is unsurprising, since both concepts are concerned with responses to stresses and perturbations. Moreover, they share a predictive quality; Cannon notes that it should be possible to identify a household's vulnerability, as well as its capacity for resilience, on the basis of its characteristics (Cannon, 2008). Vulnerability and resilience are, therefore, variously considered as being at either end of a well-being spectrum, two sides of the same coin, and part of the same equation (Cannon, 2008; Haimes, 2009; Sumner and Mallett, 2011).

While the coconut tree metaphor offers a somewhat narrow interpretation of the risk management options available to a household, it serves to illustrate that households are, to some extent, in control of their vulnerability and resilience to certain risks. In contrast to an inanimate tree, which must rely on its physical characteristics to withstand the effects of the cyclone (such as the strength of its trunk and roots, its height and age) households are able to rely on their existing asset base, as well as their capacity to deploy these assets. Indeed, according to Moser (1998) even the poorest households are managers of complex portfolios of assets (financial, human, livelihood, physical, environmental) that are mobilized and managed on a daily basis in order to protect their standard of living.

The vulnerability of a tree in a cyclone is determined by both exogenous and endogenous factors: all else being equal, the greater the exogenous risk (such as the intensity of the storm) or the weaker the tree's capacity to withstand the effects of that risk, then the more vulnerable that tree is to experiencing a fall, and vice versa. By implication, when a household is able to take actions that mitigate the risk that it faces, or is able to effectively manage the negative consequences of risk, then it is less likely to be vulnerable to suffering an adverse outcome. Therefore, in addition to assessing the extent of various risks themselves, the key challenge for policymakers interested in assisting vulnerable households is in differentiating between those factors that exacerbate risk (and hence contribute to a household's

1 An individual, community or nation can also be vulnerable to experiencing an undesirable outcome. In general, households will be the focus of this paper though, in most cases, each of these different referent objects can be directly substituted for households.

vulnerability) and those factors that help mitigate risk (and hence decrease its vulnerability, or increase its resilience).

However, an analysis of the interplay between risks and responses must also be mindful of the crucial role that the institutional setting plays in determining outcomes. For instance, broader environmental factors can either help or hinder a tree's capacity to withstand a storm's force; while strong roots may provide an effective means of resilience in a benign environment (such as when protected as part of a plantation), the same roots may prove less effective if the tree is exposed and alone on the edge of a cliff.

The important corollary of the role played by exogenous environmental factors in determining vulnerability is the role played by gender. Susman et al. identified vulnerability as 'the degree to which different classes of society are differentially at risk' (Susman et al., 1983, p. 264). Similar to the plight of the exposed tree, relatively marginalized groups in society (such as women, migrants and the disabled) often face the same risks as more powerful cohorts in society, yet are relatively more vulnerable, given their lack of control over resources and inability to mobilize them in times of need. Indeed, just as a tree can control neither soil quality nor exposure, so too are power disparities and differential access to resources often beyond the control of more marginalized groups. However, insofar as these factors are instrumental in constraining the ability to manage risk, they are determinants of vulnerability to shocks (Frankenberg and Thomas, 2003; Hoogeveen et al., 2004).

The relationship between risks and responses also has an important temporal dimension. The ultimate outcome from a given risk cannot simply be viewed from the perspective of current risks and current responses, but rather must be viewed as the cumulative effect of previous risks and responses (Alwang et al., 2001). The history of cyclones in its area, including their frequency, duration and magnitude are all crucial, determinants of whether a tree is able to withstand the next storm.

Moreover, the effects of previous risk-mitigation strategies are also likely to be important factors. For instance, if a tree has been tethered to a support this may have helped with the effects of a particular cyclone and ensured the tree remained standing during that cyclone, but if the rope and anchoring were weakened and this is not repaired the tree will probably have a much-diminished capacity to weather any subsequent storms. In a similar light, when households draw down on their asset base (such as sell livestock or withdraw children from school in order to stave off an episode of relative poverty) they may, in fact, permanently inhibit their ability to manage future risks – and thus permanently increase their vulnerability to poverty.

2.3 The Many Lenses of Vulnerability and Resilience

The concept of vulnerability provides a valuable analytical tool for describing susceptibility to harm and powerlessness and for guiding actions to enhance well-being and the reduction of risk (Adger, 2006). Therefore there are strong normative reasons why vulnerability analyses should attract the attention of policymakers.

However, there are also three key practical reasons why vulnerability should be separately identified to poverty. One is that vulnerability is both a cause and a symptom of suffering and is therefore self-reinforcing. While vulnerability is an important dimension of poverty and deprivation, highly vulnerable households may also lock themselves into low-risk and low-reward coping strategies in order to avoid the risk of an even worse outcome (Dercon, 2006). Vulnerability can thus be an important determinant of chronic poverty and give rise to poverty traps. Another key practical reason is that unlike poverty, which is a static measure of current welfare, vulnerability is a forward-looking assessment of the likelihood of suffering an unacceptable level of welfare at some point in the future. While these two concepts are related, there is no *a priori* reason why a vulnerable individual will also be poor (Dercon, 2001). Indeed, unlike poverty, the incidence and the extent of a households' vulnerability cannot, by definition, be directly observed. Instead it must be estimated using imperfect information about the future state of the world (Chaudhuri et al., 2002). To that end, a different set of tools of analysis are required, and assumptions regarding the probability distribution of future risks must be made. The third key reason is that policy measures designed to address vulnerability are likely to be quite distinct from addressing poverty. Indeed, just as vaccination is a distinctly different method for fighting the prevalence of disease to treatment of its symptoms, policies designed to prevent poverty (by decreasing risk or increasing the ability to cope with risk) are likely to be quite different to those policies designed to alleviate the current incidence of poverty (Chaudhuri, 2003).

Given the importance of focusing on vulnerability and resilience, it is unsurprising that it has become a prominent area of study across a range of different disciplines. Each makes an important contribution to the general comprehension of vulnerability, however, to date, no preferred theoretical approach, nor preferred measurement technique, has emerged. There is not, it seems, a 'gold standard'. As the raft of factors involved in determining a metaphoric coconut tree's outcome illustrates, vulnerability is a dynamic phenomenon, set against a complex broader environment with various feedback loops, making it very difficult to reduce to a single sentence or metric (Adger, 2006). The result is an inevitable trade-off between the conceptual richness of the qualitative approaches to assessing vulnerability and the precision of narrower quantitative approaches (Alwang et al., 2001). While holistic approaches can generally articulate the myriad dimensions involved in determining whether a tree falls in a cyclone, they generally do so at the expense of providing a clear measure of whether and when the tree will fall (Holzmann, 2001). However, approaches that provide a precise measure of vulnerability typically do so by abstracting from much of the rich contextual narrative. A simple taxonomy of some of the major approaches to analyzing vulnerability is presented in Table 2.1, which the remaining part of this section will more fully explain.

Table 2.1 Vulnerability: a multidisciplinary approach

Discipline	Definition	Measures	Advantages	Disadvantages
Economics	Vulnerability to falling into consumption or income poverty.	Defined relative to a minimum level of socially acceptable welfare (usually a consumption or income poverty line).	Standardized, replicable and almost universal approach.	A narrow perception of vulnerability. Excludes key non-cash aspects of vulnerability or well-being and power dynamics within or between households.
Food security	Vulnerability to falling below a critical food security threshold.	Defined relative to a benchmark level of food security (measured using human indictors such as child malnutrition, stunting and wasting).	Broadens analysis to non-cash economy and the role of nutrition in providing a sustainable livelihood.	Lacks a benchmark that is standardized and replicable. Fails to incorporate power dynamics within or between households.
Sustainable livelihoods	Vulnerability to livelihood stress.	Generally qualitative case study assessments.	A broader conceptual framework that encompasses the multidimensionality of vulnerability Incorporates livelihood assets and strategies set against the institutional environment.	The breadth of the definition makes assessing vulnerability difficult. Deep contextual aspects make comparisons difficult. Difficult to prescribe a minimum level of 'livelihood'.
Sociology	Focuses on underlying vulnerability based on social relations and power.	Usually qualitative power analyses.	Includes relational aspects of vulnerability including vulnerable groups, resource access and intra-household distribution.	Social dynamics are difficult to quantify. Deep contextual aspects make comparisons difficult.
Disaster management	Focus is on a specific type of shock – a natural hazard.	Vulnerability is a function of both the intensity of hazard and prevailing socioeconomic conditions.	Encompasses both vulnerability to the shock and the plight of the most inherently vulnerable in the context of a disaster.	Narrow focus on risks to natural disasters.

Vulnerability assessments are heavily influenced by economics – in large part because economic paradigms have tended to dominate definitions of well-being (Narayan and Petesch, 2007). The normative hold economics has over the characterization of poverty is evident in the fact that poverty lines are generally measured in monetary terms – that is the level of consumption or income needed to sustain a minimally acceptable level of well-being. Sometimes these are measured in absolute terms, such as the international poverty line of US$1.25 per day, beneath which an individual is said to suffer extreme poverty. However, even when poverty lines are re-calculated to reflect broader well-being concerns, such as the amount of money required to purchase a minimally nutritious, low-cost diet, or the cost of food plus essential non-food expenditure, they are quantified in monetary terms (ADB, 2008). While such measures of welfare tend to be reductionist and non-contextual, they nonetheless appeal to policymakers because of their simplicity, the fact they can be quantified and their comparability across time and space (Chambers, 2007).

When estimating vulnerability, economists typically utilize these quantitative poverty benchmarks, both at the microeconomic and the macroeconomic level. Looking at nations as a whole, Guillaumont (2010) asserts that some are economically vulnerable to the extent that the dual risks of natural hazards or trade and exchange-related economic shocks will hamper their economic development. In this sense, the focus is on institutional capacity to deal with the effects of exogenous shocks, with vulnerable countries those that are unable to prevent deterioration in their national income or the human development of their citizens in the face of natural and economic shocks. Two classifications have been created by the United Nations solely to account for these types of vulnerability: the Least Developed Countries (LDCs) and the Small Islands Developing States (SIDS). At the micro level, vulnerability assessments have generally examined the vulnerability of individual households to suffering a fall in well-being, with well-being typically measured in terms of the value of household income or consumption. Ironically, while economists can lay claim to having one of the more quantitatively precise approaches to measuring vulnerability, to date no definitive consensus has emerged on how to measure the concept. Authors remain divided on the utility of various approaches, including estimates of vulnerability to poverty (Pritchett *et al*,. 2000; Chaudhuri et al., 2002; Suryahadi and Sumarto, 2003; Christianesen and Subbarao, 2005; Zhang and Wan, 2006; Günther and Harttgen, 2009; Jha and Dang, 2010; Chiwaula et al., 2011), vulnerability as a shortfall in expected utility (Ligon and Schecter, 2003) and various assessments of vulnerability as uninsured exposure to risk (Glewwe and Hall 1998; Tesliuc and Lindert, 2004; Corbacho et al., 2007). However, more recently the empirical economics literature has coalesced around estimating vulnerability as the probability of experiencing poverty in the future (Zhang and Wan, 2009).

Economic approaches to assessing vulnerability are particularly useful because they present a standardized, replicable and almost universal approach that can be used to make broad comparisons of vulnerability between people in different places and time. In this sense, measuring whether a tree has fallen, or what a fall represents, is both quantifiable and standardized. Moreover, expected poverty can be calculated and disaggregated into its different causes. This is important from the perspective of implementing policies designed to decrease vulnerability or strengthen resilience. For example, it is crucial for policymakers to understand whether a household's vulnerability stems from chronic factors, such as persistently low levels of consumption, or transitory factors, such as excessive exposure to risk. Indeed, insurance designed to mitigate excessive risk exposure is unlikely to be effective in increasing the resilience of a household if its vulnerability stems from chronic poverty. Instead, targeted income support may be a more effective strategy to help that household escape poverty (Chaudhuri, 2003).

The economic approach has some obvious limitations. Because economists define vulnerability solely in terms of those variables that are measured, they tend to present a reductionist perspective of those groups that are vulnerable (Chambers, 2007). Attempts have been made to broaden the scope of vulnerability assessments beyond consumption and income, in particular to include the key role played by assets in helping households mitigate risk (Moser, 1998; Chiwaula et al., 2011). However, these generally remain predicated on being able to value these assets in monetary terms. Thus, the economic approach is ill-equipped to deal with variables for which no objective value exists – such as social, environmental and human assets – as well as immeasurable elements of social concern (including perceptions of insecurity and the role of gender in distributing resources within a household unit). While each of these contextual dimensions plays a crucial role in determining actual vulnerability, the inherent difficulty of accounting for these broader concerns means that they tend to be either ignored or subsumed into the economic model via assumptions regarding their relationship with consumption or income. Accordingly, while economists can produce precise measures of vulnerability that are comparable across time and space there are limits to what is actually conveyed. A parallel might be to base an assessment of the vulnerability of all plants (including delicate flowers, saplings and sturdy trees) in a given location to the effects of a cyclone by only precisely measuring the vulnerability of coconut trees.

Similar to the economic approach, the food security literature also analyses whether a household is vulnerable to experiencing an unacceptable standard of welfare at some point in the future (Scaramozzino, 2006). However, rather than being defined in monetary terms, welfare is instead characterized in terms of having sufficient nutrients to consume. For example, the UN Food and Agricultural Organization (FAO) offered a definition of food security at the 1996 World Food Summit Plan of Action, which was the requirement that a household have 'physical, social and economic access to sufficient safe and nutritious food to meet their dietary needs and their food preferences to meet lead a healthy life' (FAO, 1996).

To the extent that both concentrate on households' resilience to various shocks in order to sustain a minimally acceptable standard of living, economic and food security approaches have similar conceptual frameworks. However, while economists concentrate on the overall material well-being of a household, food security analysts explicitly focus on the dimensions of food security, including availability, access and utilization, each of which is a necessary but not a sufficient condition for food security, and thus well-being (Maxwell and Smith, 1992; Pinstrup-Andersen, 2009; Barrett, 2010).

One could think of this difference between economics and food security as the difference between basing the overall vulnerability of the tree on how it appears above the ground as distinct from focusing mainly on less-visible aspects of the tree's health, such as its root structure and ability to source nutrients from the soil. While on the face of it a tree may appear strong, if in reality its roots are eroded or its core is brittle then, in all likelihood, its vulnerability is likely to be heightened. Accordingly, some economists have taken to using indicators of food intake, such as per capita food consumption as the relevant indicator of wellbeing in their models of vulnerability (Kühl, 2003; Kamanou and Morduch, 2004) while others consider vulnerability in terms of the probability of being undernourished in the future (Christiaensen and Boisvert, 2000).

However, the economics and food security approaches are not direct substitutes. Unlike economics, food security approaches to vulnerability lack a consistent and quantifiable measure of exactly what food security looks like. Because an individual's food intake is a function of a multitude of factors, there is no single measure that encapsulates all the aspects of food security (Scarramozzino, 2006). Without such a benchmark, identifying exactly what households are vulnerable to is subjective and open to interpretation. While studies have typically employed a variety of anthropometric instrumental indicators that reflect the effects of food insecurity (such as child malnutrition, stunting and wasting) each of these are affected by much more than simply the security of food supplies. Indeed, a household can be food insecure yet not be experiencing hunger (Webb et al., 2006; FAO, 2011). Thus comparisons of vulnerability to experiencing food security across time and space are difficult. Moreover, as is the case with economics, using food security as the relevant lens of household vulnerability tends to say little about the roles that gender and power play in the intra-household distribution of food. It also provides limited information about the specific mechanisms households use to maintain their caloric intake during shocks (such as through substitution of lower quality food in times of food price inflation) and how this may affect their ongoing well-being.

The sustainable livelihoods approach, in some respects, fills gaps in both the economics and food security perspectives by placing vulnerability and resilience in the context of a household's connections with its broader environment. As the name suggests, the approach considers those factors that impact upon the ability of individuals and groups to sustain a livelihood. In this context 'livelihoods', can range from the 'means of generating a living', to the 'combination of resources used and activities undertaken in order to live' (Chambers and Conway, 1992, p. 5). Rather than attempting to construct a measure of vulnerability *per se*, the sustainable livelihoods approach is more akin to a way of thinking about the broader contextual aspects of people's circumstances. To that end, a focus on 'livelihoods' (a deliberately nebulous term), encapsulates the myriad decisions that households make regarding their activities and resource allocations and the complex way these interact with underlying political, economic and social processes to comprise a households' way of subsisting (Scoones, 2009). A 'sustainable livelihood', then, is one that can cope with, and recover from, stress and shocks, while not undermining its resource base, in order to provide opportunities for the next generation (Chambers and Conway, 1992).

The sustainable livelihoods approach is therefore a unique and important analytical tool. Assessments of vulnerability concentrate on the root causes of vulnerability, that is the institutional factors that weaken the capacity of a household to adapt and cope with risk going forward (Cannon et al., 2003). Importantly, because vulnerability is characterized as the general insecurity of the well-being of households in the face of environmental changes, vulnerability assessments appropriately recognize the future implications of coping mechanisms. Just as a tree that sustains damage while withstanding the effects of a storm may be vulnerable to future storms, the sustainable livelihoods approach not only focuses on a household's ability to cope with current risks, but also the longer-term impacts of various coping strategies and how these actions affect its welfare over time (Serrat, 2008).

The conceptual strength of the sustainable livelihoods approach is tempered somewhat by its lack of a quantifiable benchmark. Livelihood analyses can generally only be presented as qualitative case studies (Janssen and Ostrom, 2006). Such analyses are necessarily subjective and provide little guidance as to what is a minimum acceptable level of livelihood, or a point at which livelihood stress occurs. Moreover, they are problematic when comparisons are required between different case studies – as often occurs when deciding on how to allocate scarce resources. To that end, sustainable livelihoods approaches are likely to have greater utility in providing background and contextual insights rather than in estimating vulnerability *per se*. Indeed, while the focus on sustainable livelihoods can comprehensively articulate the reasons why a tree may be vulnerable in a cyclone, it does not convey exactly what vulnerability in each case represents – indeed, is a tree vulnerable to falling over in a cyclone or vulnerable to simply developing a lean?

Similar to the sustainable livelihoods approach, sociology broadens the analysis of vulnerability beyond the physical and financial realms. Sociology introduces inherently human aspects, including social capital, capabilities, access to resources and human agency (Loughhead and Mittai, 2000). In this light, vulnerability is framed in terms of social vulnerabilities, which have at their core social relations, as well as the insecurity of the well-being of particular individuals, households or communities in the face of a changing environment (Downing et al., 2006). To the extent that social networks are one of the primary resources the poor have at their disposal to manage risk, communities relatively endowed with a diverse stock of social capital are likely to be in a stronger position to confront vulnerability *vis-à-vis* a less well-endowed community (Woolcock and Narayan, 2000). Accordingly, the sociological approach identifies those groups that are inherently more vulnerable because of broader institutional factors beyond their control – not their own characteristics (Soussan et al., 2001; Hoogenveen et al., 2004). Vulnerability assessments, therefore, focus less on the risk-response–outcome relationship between the tree and the cyclone, and more on the factors, beyond the tree's control, that either help or hinder its ability to deal with risk.

By broadening the perspective of vulnerability beyond narrow materialistic interpretations of well-being, sociology provides insights into vulnerability that may otherwise be ignored by approaches, such as economics and food security that focus on instrumental indicators. Sociology's unique contribution to vulnerability analysis is in viewing an individual's vulnerability through the lens of the choices available. To that end, a focus is given to the importance of power relations, both within and across households, across communities and between genders, each of which exerts some influence over the control people have over their resources and their ability to manage risk (Horn, 2010).

Sociology, therefore, provides a complement to many of the other perspectives on vulnerability. For example, sociology drills deeper into economic assessments of household vulnerability to analyze the role of women in the domestic environment (an area that is typically not measured and often ignored).

However, as with sustainable livelihoods, the drawback to the sociological approach is in its lack of a quantifiable benchmark. Sociological estimates of vulnerability necessarily reflect unique and contextual factors. This means that making robust assessments of who is in fact vulnerable and comparisons of different degrees of vulnerability are particularly problematic. Consequently, such assessments are unlikely to be easily subsumed into existing approaches. Indeed, just as an assessment of the broader environmental factors assisting or hindering a tree's ability to cope with risk requires a broader knowledge of the environment (such as soil types and weather patterns), comprehending the prevailing social dynamics of a region requires a deep ethnographic understanding of the history, culture and practices of individual communities. Notwithstanding these practical difficulties, there is a view – not without merit – that the benefits of including sociological perspectives in existing vulnerability analyses will almost always outweigh their costs (Chambers, 2007).

The disaster management literature takes a somewhat different perspective on the concept of vulnerability by turning attention to the nature of the risk itself. Disaster management literature frames the analysis of vulnerability more in terms of the cyclone (literally, in some cases) rather than the coconut tree (Yamin et al., 2005). Vulnerability is a core concept of disaster risk and the common approach in the literature is to focus on sensitivity to the negative effects of a particular shock. The International Panel on Climate Change, for instance, considers vulnerability as 'the extent to which a natural or social system is susceptible to sustaining damage from climate change. Under this framework, a highly vulnerable system would be one that is highly sensitive to modest changes in climate' (IPCC 2001,

p. 89). Cannon conceptualizes this even more anthropocentrically, noting that, 'natural disasters are not in themselves natural: a disaster only happens when a hazard has an impact on vulnerable population' (Cannon, 2008, p. 2). Thus, it is recognized explicitly that a given household's vulnerability will be determined by the characteristics of a given natural hazard, as well as some combination of factors that either mitigate or accentuate its effects, including, for instance, pre-existing economic conditions, control over resources or simply luck (Blaikie et al., 1994).

The disaster management literature, therefore, intersects neatly with each of the other approaches to vulnerability. Like most other approaches, vulnerability is viewed as a function of the intensity of a hazard and the ability of households to mitigate this risk to prevent an adverse outcome. However, by keeping the hazard/shock at the centre of the analysis each of the main approaches to vulnerability are considered in the analysis of the potential outcomes. For instance, authors note while a hazard (such as a cyclone) may be experienced by all households in a given location, it is those households that are already poor, that live on marginal land, and those that lack the wherewithal to access disaster relief resources that are likely to be most vulnerable to the adverse effects of the shock (Douglas, 2009; Winchester and Szalachman, 2009).

While little can be done to prevent natural disasters (the shocks themselves), the advantage of this approach is that it focuses on the raft of factors that either exacerbate of mitigate a household's exposure to a risk. It therefore provides a useful template for policymakers looking to analyze vulnerability to human-induced hazards, such as adverse macroeconomic shocks.

The policy focus of vulnerability-to-disaster analysis is on strengthening resilience and the ability to withstand shocks. By blending those dimensions of vulnerability that are quantifiable with those that are less tangible (though no less important) policymakers can establish a clearer sense of the various factors involved in determining vulnerability and resilience. Such a comprehensive perspective is likely to yield more effective policy decisions than when only partial information – either entirely quantitative or qualitative – is available.

2.4 Vulnerability and Resilience in Melanesia

The small island states of Melanesia are inherently vulnerable to the effects of a variety of natural and market-induced exogenous shocks. Situated in the tropics, and straddling the Pacific 'Ring of Fire', these states are highly vulnerable to the effects of disasters such as earthquakes and cyclones (Guillaumont, 2010). On the economic front, the combination of smallness and remoteness means they also face severe economic disadvantages, including a small domestic resource base, a lack of large-scale economies and small domestic markets (Gibson and Nero, 2008). Such intrinsic disadvantages uniquely hinder the economic development prospects of Pacific Island economies (Jayasuriya and Suri, 2012). To the extent that such smallness and remoteness necessitates a high degree of economic openness, these small island states, with a narrow export base and a dependence on strategic imports, are also highly exposed to terms-of-trade shocks and fluctuations in international economic activity (Jayaraman, 2004; Briguglio, 2011).

Such high exposure to exogenous shocks, coupled with limited institutional capacity to deal with them, helps explain why a number of international tables place two Melanesian countries, the Solomon Islands and Vanuatu, among the world's most vulnerable countries. The United Nations University Institute 2011 World Risk Report ranks Vanuatu and the Solomon Islands as the world's first and fourth nations most vulnerable to natural hazards, while both countries rank among the world's 11 most vulnerable countries, according to the UN Committee for Development Policy's Economic Vulnerability Indicators (EVI).

Crucially, however, exposure to shocks is only one side of the vulnerability story. Key dimensions of resilience in Melanesia provide insurance against the effects of external risks. Feeny identifies four buffers that Melanesian countries possess that insulated them from the effects of the global financial crisis: (i) limited financial integration with global financial markets; (ii) the dominance of communally owned land and strong traditional social support systems; (iii) low levels of monetization and high rates of subsistence agriculture; and (iv) low levels of formal sector employment (Feeny, 2010). Indeed, the dominance of traditional livelihoods, such as customary dispute resolution, universal access to land on which to access food and make a living, and strong familial ties characterized by norms of sharing and reciprocity, called the 'traditional economy' (Regenvanu 2009, p. 30), provides households in Melanesia with a crucial resilience mechanism: by thwarting the transmission of externally generated macroeconomic shocks as well as providing a social security system that is largely absent in a formal sense (see Chapter 6). Yet, despite being such a dominant influence on livelihoods, as well as an important source of value-add, the underlying strength of the traditional economy is often not visible in more formal economic analyses. Indeed, smallholder subsistence farming systems have represented a 'hidden strength of otherwise structurally weak economies' in Melanesia given their capacity to provide access to nutritional food as well as an informal safety net in the absence of any substantive formal arrangements (McGregor et al., 2009, p. 26).

The traditional economy is increasingly coming under strain. Rapid rates of urbanization in both countries, coupled with the gradual monetization of rural areas, are underpinning a shift away from subsistence agriculture and towards imported food (Pacific Institute of Public Policy 2011a; 2011b). The upshot is that as households turn toward urban livelihoods and the trappings of modernity they are becoming alienated from the land. The increased exposure to the vagaries of international food and fuel markets, in turn, is increasing households' vulnerability to exogenously determined price shocks, such as the rapid inflation in food and fuel prices in recent years. Moreover, urbanization and economic development are typically accompanied by increased commodification of labour and greater social and economic heterogeneity (Moser, 1998). To the extent that these shifts can have deleterious effects on informal safety nets households may well be becoming more vulnerable, owing to the erosion of a key resilience mechanism.

These social shifts are redefining the role of customary practices in modern Melanesia. Consequently, it is understandable that such developments are among the most predominant issues raised in focus group discussions in both the Solomon Islands and Vanuatu as part of an investigation into household vulnerability and resilience. In urban and rural communities, and across the gender divide, the key theme to emerge is the gradual social fragmentation that has accompanied increased urbanization and monetization; and the adverse effect this has had on the underlying social fabric. According to the information gathered in focus groups, households are facing a combination of rising costs of living, as well as inadequate employment opportunities and low wages. With no formal safety net to speak of, communities continue to rely on the traditional informal support practices, built on a long-standing sense of obligation and trust. However, these ties appear to be under threat, as the increased sense of desperation faced by many households is spilling over, at least at the margin, to increased rates of theft and even substance abuse. To the extent that these outcomes have potentially deleterious implications for social cohesion and the ongoing integrity of social links, they threaten to even further exacerbate households' vulnerability to exogenous shocks.

2.5 Conclusion

The World Bank articulated in its 2000–2001 World Development Report that sustainable poverty reduction required a forward-looking approach to reducing vulnerability (World Bank, 2001). Since then it has become increasingly evident that preventing poverty, through strengthening the ability of households, regions and countries to cope with risk, is just as important as alleviating it. Consequently, vulnerability assessments, which seek to determine who is vulnerable and to what, are rapidly becoming a cornerstone of development policy.

Yet providing an overarching definition of vulnerability is not straightforward. While vulnerability assessments are a feature of a number of different disciplines, there remains no consensus regarding the definition of the term, nor an agreed-upon estimation method. To be sure, some of this definitional imprecision is likely to be caused by the intellectual fragmentation of vulnerability; yet even within disciplines, such as economics, there is little consensus on how to conceptualize, or even estimate, vulnerability.

The proliferation of approaches to analyzing vulnerability is nonetheless instructive. Vulnerability (and by implication resilience) is multidimensional: it encompasses both extrinsic and intrinsic aspects and is dynamic, in the sense that actions taken today will have consequences tomorrow. It also reflects important social considerations such as the ability to mobilize resources. Many of these dimensions are unobservable; two households ostensibly in the same situation may nonetheless be differentially vulnerable. The upshot is that there is no benchmark and thus vulnerability assessments typically involve a trade-off between precision and breadth.

However, precision and breadth need not be mutually exclusive when analyzing vulnerability. Some optimal mixture of quantitative and qualitative information is likely to furnish policymakers with sufficient information to make genuinely informed decisions regarding the prevention of poverty. Melanesia provides a particularly interesting case in point. Traditional mechanisms of informal social insurance, that rely on complex and unobservable social interactions, are as fundamental to the ability of households in the Solomon Islands and Vanuatu to withstand the effects of exogenous shocks as strong roots are to a tree's chances of negotiating a cyclone. Yet they risk being ignored altogether if vulnerability assessments are based solely on those characteristics that are both observable and measureable. Such characterizations of vulnerability are likely to be blind to the underlying resilience and strength exhibited by Melanesian communities in the form of the traditional economy. Accordingly, they will also be unlikely to appreciate the magnitude of the social shifts that are underway in Melanesian society and the extent to which they are exerting a substantial influence on the vulnerability of households to experiencing poverty. Indeed, just as a coconut tree's vulnerability to the effects of a cyclone is the function of a multitude of factors, which cannot be determined by a single measure alone, the true essence of a households' vulnerability to poverty is likely to be appreciated only when the precision of quantitative data is able to be placed in an appropriate context by qualitative explanations of unobservable phenomena. More specifically, while economics appears to have a normative hold on how welfare is measured, the empirical rigor of economic approaches to assessing vulnerability to poverty are likely to be substantially strengthened if they are combined with qualitative dimensions that drill deeper into the non-monetary space.

References

ADB (2008), *Vanuatu: Analysis of the 2006 Household Income and Expenditure Survey; A Report on the Estimation of Basic Needs Poverty Lines and the Incidence and Characteristics of Poverty in Vanuatu* (Asian Development Bank: Manila).

Adger, N.W. (2006), Vulnerability, *Global Environmental Change,* 16(3): 268–81.

Alwang, J., Siegel, P.B. and Jorgensen, S. (2001), Vulnerability: A View from Different Disciplines, *Social Protection* (World Bank: Washington).

Barrett, C.B. (2010), Measuring Food Insecurity, *Science*, 327(5967): 825–8

Blaikie, P., Cannon, T., Davis, I. and Wisner, B. (1994), *At Risk: Natural Hazards, People's Vulnerability, and Disasters* (Routledge: London).

Briguglio, L. (2011), Economic Vulnerability and Resilience with Reference to Small States, *Presentation to First meeting of the Caribbean Development*

Round Table (September 13, 2011) (The Economic Commission for Latin America (ECLA): Port of Spain).

Cannon, T. (2008), Reducing People's Vulnerability to Natural Hazards: Communities and Resilience, *United Nations University Research Paper No. 2008/34* (World Institute for Development Economics Research: Bonn).

Cannon, T., Twigg, J. and Rowell, J. (2003), *Social Vulnerability, Sustainable Livelihoods and Disasters* (Department for International Development: London).

Chambers, R. (2006), Vulnerability, Coping and Policy (Editorial Introduction), *Institute of Development Studies Bulletin,* 37(4): 33–40.

Chambers, R. (2007), Poverty Research: Methodologies, Mindsets and Multidimensionality, *Institute of Development Studies Working Paper 293* (Institute of Development Studies: Brighton).

Chambers, R. and Conway, G. (1992), Sustainable Rural Livelihoods: Practical Concepts for the 21st Century, *IDS Discussion paper 296* (Institute of Development Studies: Brighton)

Chaudhuri, S. (2003), *Assessing Vulnerability to Poverty: Concepts Empirical Methods and Illustrative Examples* (Columbia University Department of Economics: New York).

Chaudhuri, S., Jalan, J. and Suryahadi, A. (2002), Assessing Household Vulnerability to Poverty from Cross Sectional Data: A Methodology and Estimates from Indonesia. *Discussion Paper No. 01022-52* (Columbia University: New York).

Chiwaula, L.S., Witt, R. and Waibel, H. (2011), An Asset-based Approach to Vulnerability: The Case of Small-Scale Fishing Areas in Cameroon and Nigeria, *The Journal of Development Studies,* 47(2): 338–53.

Christiaensen, L. and Boisvert, R. (2000), On Measuring Household Food Vulnerability: Case Evidence from Northern Mali, *Working Paper 2000–2005* (Department of Applied Economics and Management Cornell University: New York).

Christiaensen, L. and Subbarao, K. (2005), Towards an Understanding of Household Vulnerability in Rural Kenya, *Journal of African Economies,* 14(4): 520–58.

Corbacho, A., Garcia-Escribano, M. and Inchaustel, G. (2007), Argentina: Macroeconomic Crisis and Household Vulnerability, *Review of Development Economics,* 11(1): 92–106.

Dercon, S. (2001), *Assessing Vulnerability to Poverty* (Department of Economics, University of Oxford: Oxford).

Dercon, S. (2006), Vulnerability: A Micro Perspective, *QEH Working Paper Series – QEHWPS149* (Department of International Development, University of Oxford: Oxford).

Douglas, I. (2009), Climate Change, Flooding and Food Security in South Asia, *Food Security,* 1: 127–36.

Downing, T.E., Aerts, J., Soussan, J., Barthelemy, O., Bharwani, S., Hinkel, J., Ionescu, C., Klein, R.J.T., Mata, L.J., Matin, N., Moss, S., Purkey, D. and Ziervogel, G. (2006), Integrating Social Vulnerability into Water Management, *SEI Working Paper and Newater Working Paper No. 4* (Stockholm Environment Institute: Stockholm).

FAO (1996), Declaration on World Food Security, *World Food Summit*, Food and Agricultural Organization, Rome.

FAO (2011), The State of Food Insecurity in the World 2011, Food and Agricultural Organization, Rome. [accessed 23 May 2012], http://www.fao.org/docrep/014/i2330e/i2330e00.htm

Feeny, S. (2010), *The Impact of the Global Economic Crisis on the Pacific Region*. (Oxfam Australia: Melbourne).

Frankenberg, E. and Thomas, D. (2003), Measuring Power, in A.R. Quisumbing (ed.), *Household Decisions, Gender, and Development: A Synthesis of Recent Research* (International Food Policy Research Institute, Johns Hopkins University Press: Baltimore).

Gibson, J. and Nero, K. (2008), Why Don't Pacific Island Countries' Economies Grow Faster?, in A. Bisley (ed.), *Pacific Interactions: Pasifika in New Zealand – New Zealand in Pasifika* (Institute of Policy Studies, Victoria University of Wellington: Wellington).

Glewwe, P. and Hall, G. (1998), Are Some Groups More Vulnerable to Macroeconomic Shocks Than Others? Hypothesis Tests Based on Panel Data from Peru, *Journal of Development Economics,* 56(1): 181–206.

Guillaumont, P. (2010), Assessing the Economic Vulnerability of Small Island Developing States and the Least Developed Countries, *Journal of Development Studies,* 46(5): 828–54.

Günther, I. and Harttgen, K. (2009), Estimating Households Vulnerability to Idiosyncratic and Covariate Shocks: A Novel Method Applied in Madagascar, *World Development,* 37(7): 1222–34.

Haimes, Y.Y. (2009), On the Definition of Resilience in Systems, *Risk Analysis,* 29(4): 498–501.

Heitzmann, K.R., Canagarajah, S. and Siegel, P. (2002), Guidelines for Assessing the Sources of Risk and Vulnerability, *Social Protection Discussion Paper Series, No. 0218* (World Bank: Washington).

Hoddinott, J. and Quisumbing, A. (2003), Methods for Microeconometric Risk and Vulnerability Assessment, *Social Protection Discussion Paper Series, No. 0324* (World Bank: Washington).

Holzmann, R. (2001), Risk and Vulnerability: the Forward Looking Role of Social Protection in a Globalizing World, *Social Protection Discussion Papers 0324* (World Bank: Washington).

Hoogeveen, J, Tesliuc, E., Vakis, R. and Dercon, S. (2004), *A Guide to the Analysis of Risk, Vulnerability and Vulnerable Groups* (World Bank and University of Oxford: Washington).

Horn, Z. (2010), No Cushion to Fall Back On: The Impact of the Global Recession on Women in the Informal Economy in Four Asian Countries, *Working Paper for Conference 'The Impact of the Global Economic Slowdown on Poverty and Sustainable Development in Asia and the Pacific'*, 28–30 September 2009 (Hanoi, Asian Development Bank: Manila).

IPCC (2001), *Climate Change 2001: Impacts, Adaptation, and Vulnerability, Contribution of Working Group II to the Third Assessment Report of the Intergovernmental Panel on Climate Change* (Intergovernmental Panel on Climate Change: Cambridge University Press, UK.

Janssen, M.A. and Ostrom, E. (2006), Resilience, Vulnerability, and Adaptation: A Cross-Cutting Theme of the International Human Dimensions Programme on Global Environmental Change. *Global Environmental Change,* 16(3): 237–9.

Jayaraman, T.K. (2004), Coping with Vulnerability by Building Economic Resilience: The Case of Vanuatu, in L. Briguglio and L. Eliawony (eds), *Economic Vulnerability and Resilience of Small States* (Islands and Small States Institute, University of Malta: Malta).

Jayasuriya, D. and Suri, V. (2012), Impact of Smallness and Remoteness on Growth: The Special Case of the Pacific Island Countries, *ANU Working Paper* (Australian National University: Canberra).

Jha, R, and Dang T. (2010), Vulnerability to Poverty in Papua New Guinea in 1996, *Asian Economic Journal,* 24(3): 235–51.

Kamanou, G. and Morduch, J. (2004), Measuring Vulnerability to Poverty, in S. Dercon (ed.) *Insurance Against Poverty* (Oxford University Press: Oxford).

Kühl, J.J. (2003), *Disaggregating Household Vulnerability-Analyzing Fluctuations in Consumption Using a Simulation Approach*, mimeo (Institute of Economics, University of Copenhagen: Copenhagen).

Ligon, E. and Schechter, L. (2003), Measuring Vulnerability, *Economic Journal,* 113(486): C95–C102.

Loughhead, S. and Mittai, O. (2000), Urban Poverty and Vulnerability in India: A Social Perspective, *Paper Presented for the Urban Forum: Urban Poverty Reduction in the 21st Century, April 3–5, 2000*, sponsored by the World Bank, Chantilly, Virginia.

McGregor, A.R., Bourke, M., Manley, M., Tubuna, S. and Deo, R. (2009), Pacific Islands Food Security: Situation Challenges and Opportunities, *Pacific Economic Bulletin,* 24(2): 24–42.

Maxwell, S. and Smith, M. (1992), Household Food Security: A Conceptual Review, in S. Maxwell and T. Frankenberger (eds), *Household Food Security: Concepts, Indicators, Measurements: A Technical Review* (UNICEF and IFAD: New York and Rome).

Miller, F., Osbahr, H., Boyd, E., Thomalla, F., Bharwani, S., Ziervogel, G., Walker, B., Birkmann, J., van der Leeuw, S., Rockström, J., Hinkel, J., Downing, T., Folke, C. and Nelson, D. (2010), Resilience and Vulnerability: Complementary or Conflicting Concepts? *Ecology and Society,* 15(3): 11.

Moser, C.N. (1998), The Asset Vulnerability Framework: Reassessing Urban Poverty Reduction Strategies, *World Development,* 26(1): 1–19.

Narayan, D. and Petesch, P. (2007), Agency, Opportunity Structure and Poverty Escapes, in D. Narayan and P. Petesch (eds), *Moving Out of Poverty* (World Bank: Washington).

Naudé, W., Santos-Paulino, A. and McGillivray, M. (2009), Measuring Vulnerability: An Overview and Introduction, *Oxford Development Studies,* 37(3): 183–91.

Pacific Institute of Public Policy (2011a), Urban Hymns: Managing Urban Growth in the Pacific, Discussion Paper 18, July 2011. [accessed 30 May 2012] http://www.pacificpolicy.org/wp-content/uploads/2012/05/D18-PiPP.pdf.

Pacific Institute of Public Policy (2011b), Food for Thought: Exploring Food Security in the Pacific, *Discussion Paper 19*, December 2011 [accessed 30 May 2012] http://www.pacificpolicy.org/wp-content/uploads/2012/05/D19-PiPP.pdf

Pinstrup-Andersen, P. (2009), Food security: Definition and Measurement, *Food Security*, 1(1): 5–7.

Pritchett, L., Suryahadi, A., and Sumarto, S. (2000), Quantifying Vulnerability to Poverty: A Proposed Measure with Applications to Indonesia, SMERU Working Paper, (Social Monitoring and Early Response Unit, World Bank: Washington).

Regenvanu, R. (2009), *The Traditional Economy as the Source of Resilience in Melanesia* (Vanuatu Cultural Centre: Port Vila).

Scaramozzino, P. (2006), Measuring Vulnerability to Food Insecurity, ESA Working Paper No. 06-12 (Agricultural and Development Economics Division, Food and Agriculture Organization: Rome).

Scoones, I. (2009), Livelihoods Perspectives and Rural Development, *The Journal of Peasant Studies,* 36(1): 171–96.

Serrat, O. (2008), The Sustainable Livelihood Approach, Knowledge Solutions 15, Asian Development Bank, Manila, [accessed 29 May 2012] http://www.adb.org/Documents/Information/Knowledge-Solutions/Sustainable-Livelihoods-Approach.pdf.

Soussan, J., Blaikiem, P., Springate-Baginski, O. and Chadwick, M. (2001), Understanding Livelihood Processes and Dynamics, *Livelihood-Policy Relationship in South Asia Working Paper* 1, University of Leeds, UK.

Sumner, A. and Mallett, R. (2011), Snakes and Ladders, Buffers and Passports: Rethinking Poverty, Vulnerability and Wellbeing. *International Policy Centre for Inclusive Growth Working Paper Number 83*, [accessed 5 February 2012] http://www.ipc-undp.org/pub/IPCWorkingPaper83.pdf.

Suryahadi, A. and Sumarto, S. (2003), Poverty and Vulnerability in Indonesia Before and After the Economic Crisis, *Asian Economic Journal,* 17(1): 45–64.

Susman, P., O'Keefe, P. and Wisner, B. (1983), Global Disasters: A Radical Interpretation, in K. Hewitt (ed.), *Interpretations of Calamity from the Viewpoint of Human Ecology* (Allen & Unwin: Boston).

Tesliuc, E. and Lindert, K. (2004), Risk and Vulnerability in Guatemala: A Quantitative and Qualitative Assessment, *Social Protection Discussion Paper 0404* (World Bank: Washington).

Webb, P., Coates, J., Frongillo, E.A., Rogers, B.L., Swindale, A. and Bilinsky, P. (2006), Measuring Household Food Insecurity: Why It's So Important and Yet So Difficult to Do?, *The Journal of Nutrition*, 136(5): 1404S–408S.

Winchester, L. and Szalachman, R. (2009), The Urban Poor's Vulnerability to the Impacts of Climate Change in Latin America and the Caribbean – A Policy Agenda, Paper prepared for Fifth Urban Research Symposium 2009, (World Bank: Washington).

Woolcock, M. and Narayan, D. (2000), Social Capital: Implications for Development Theory, Research, and Policy, *World Bank Research Observer,* 15(2): 225–49.

World Bank (2001), *World Development Report 2000/01*, World Bank, Washington.

Yamin, F., Rahman, A. and Huq, S. (2005), Vulnerability, Adaptation and Climate Disasters, A Conceptual Overview, *IDS Bulletin*, 36 (Institute of Development Studies: Brighton).

Zhang, Y. and Wan, G. (2006), An Empirical Analysis of Household Vulnerability in Rural China, *Journal of the Asia Pacific Economy*, 11(2): 196–212.

Zhang, Y. and Wan, G. (2009), How Precisely Can We Estimate Vulnerability to Poverty? *Oxford Development Studies,* 37(3): 277–87.

Chapter 3
Responding to Shocks: Women's Experiences of Economic Shocks in the Solomon Islands and Vanuatu

Jaclyn Donahue, Kate Eccles and May Miller-Dawkins

3.1 Introduction

Since 2007, the world economy has experienced a fiercely cyclical phase, punctuated by a number of major shocks. A surge in food and fuel prices through 2007 and 2008 tended to make poor households poorer (World Bank, 2008). A financial crisis emanating from the United States manifested itself into the Global Economic Crisis (GEC) with an accompanying contraction in global demand and trade. Impacts were felt acutely in many developing countries through collapsing demand for their trade-exposed manufactures, the producers of which employ women in disproportion. Even though more men in the United States and Europe experienced job loss as a result of the GEC (Otobe, 2011), in many countries women were likely to lose their jobs due to employment in precarious sectors or discrimination that sees them fired first (Parks et al., 2009; Green et al., 2010; King and Sweetman, 2010; Seguino, 2010). In Cambodia, for instance, where women are largely employed by the highly trade-exposed textile and clothing industries, women were most affected by over 38,000 industry job losses in 2009 (Otobe, 2011). In Nicaragua, when a textile factory closed in 2009, over 85 per cent of jobs lost were held by women (Silva, 2009).

In contrast, loss of labour-intensive, export-dependent employment had less impact upon the Pacific Islands due to low levels of formal sector employment (Green et al., 2010). Women conduct about 90 per cent of fresh produce marketing in the Solomon Islands (Hedditch and Manuel, 2010a), and only one-third of formal sector workers are women in Vanuatu (Vanuatu Statistics Office, 2000 as cited in Hedditch and Manuel, 2010b). Consequently, women in these countries were less likely to have been directly exposed to the effects of the GEC, by way of trade and manufacturing, as they were elsewhere. However, the research in this chapter finds that households' reactions to shocks in the Solomon Islands and Vanuatu, similar to those in other developing country settings, possess a gendered dimension as women, in particular, have strived to maintain a minimum level of well-being for their families. This is consistent with the notion that women frequently respond as 'shock absorbers' during crises by taking on additional work, which can result in significant 'time poverty' (Mendoza, 2009, p. 65, 66; Parks et al., 2009). The level of unpaid care work and responsibilities to the household continue for women alongside their potentially increasing hours of external work as they seek to supplement the family income (Horn, 2011; Heltberg et al., 2012).

The objective of this chapter is to understand whether recent shocks and subsequent responses to mitigate the impact of shocks can be understood as 'gendered' in their effect on women in the Solomon Islands and Vanuatu. It concludes that the shocks were 'gendered' insofar as women responded in particular ways to the economic, food and fuel crises. These responses are often reflective of gender roles, responsibilities and inequalities. In the Pacific region, gender inequality is evidenced in women's difficulty in entering formal sector employment and earning an income, a relative lack of access to decision-making, and to productive assets and services, responsibility for unpaid work and experiences of violence (World Bank, 2012). The data from this research reflects the observation of King and Sweetman (2010, p. 4): 'The gender inequalities and power imbalances that predate the current crisis have resulted in its additional afflictions falling disproportionately on those who are already structurally disempowered and marginalized'.

In consideration of these observations, this chapter examines the impacts of the recent economic, food and fuel crises on women in the Solomon Islands and Vanuatu and presents the findings from extensive fieldwork conducted in these countries. The research paid special attention to the gendered experiences of economic shocks through its quantitative and qualitative methodology, and analysis of the data demonstrates that women in the Solomon Islands and Vanuatu took on a distinct burden. They continued unpaid care work for the family and home with the added pressure of generating income and managing rising costs in an attempt to maintain household resilience. Unlike in some other developing countries, the GEC did not have the same impact on formal employment of Solomon Islander and ni-Vanuatu women on account of small trade and manufacturing sectors and women's underrepresentation in formal sector employment.

Section 3.2 of the chapter details the methods used to gain insight into the particular experiences of women, men, young women and young men in the research. After reviewing the findings of the quantitative and qualitative data in Section 3.3, the results are discussed against the existing literature on gender roles and inequities in Melanesia and the international literature on gender and economic shocks in Section 3.4. Finally, Section 3.5 concludes with a brief discussion addressing gender inequality in policy, economy and development.

3.2 Methodology

The methodology for the research recognizes the potentially different experiences of, and responses to, shocks by women, men, young women and young men. To collect both quantitative and qualitative data, men and women from the Solomon

Islands and Vanuatu were recruited to form a research team within their respective countries. A mixed-methods approach was used in fieldwork, consisting of the administration of more than 1,000 household surveys across six locations in each country as well as more than 50 focus group discussions and a small number of key informant interviews (see Chapter 1 for further details).

The unique household survey sought to capture a broad suite of information on well-being in Melanesia. In addition to monetary measures of income and consumption, information was collected on the demographic characteristics of the household, its asset base (including social relations) and indicators of food security. Information was also gathered on households' experiences of various economic and other shocks as well as the way that households coped with such shocks. A number of survey questions also specifically targeted intra-household gender relations by asking whether women or men did particular types of work and had access to particular household resources.

The household survey was designed to capture the diversity of experiences between men and women in three important ways. Firstly, its approach did not rely on a notion of the 'head of the household' as a respondent. In addition to the practical difficulties of sampling the household head, feminist critiques of 'household head' surveys highlight the fact that they mask intra-household dynamics and privilege the perspectives and insights of one member (Monk and Hanson, 1982). Moreover, in-country sources suggested that the 'household head' is not a useful reference point in Melanesia – a view confirmed during the piloting of surveys: the nominated 'head' of the house was typically indicated as the oldest male in a family (irrespective of whether that person resided in the house or not). This suggests that the concept of the 'head' may be simply titular in Melanesia and might not have any meaningful correlation with the economic decision-making of the household.[1] Consequently, the household survey collected information from 'the people living in this house' rather than the household head. The starting point was that if any one individual in the house had access to a particular asset, then all members of the house also, at least vicariously, had access to that asset.

Secondly, the sampling aimed for a 45 to 55 per cent gender balance in survey respondents who commented on the circumstances of the house in which they were living. The data from the surveys can therefore be analysed and contrasted between the gender of the reported head of the household and, separately, between respondents of different genders.

Thirdly, in order to contrast the gendered experiences within individual households, about 6 per cent of houses were surveyed twice: one survey of a man within the house and one of a woman. These double surveyed households allow for analysis of different perceptions and experiences within a single household.

1 The survey responses were able to illustrate the fluidity of the 'head' of the household in Melanesia. Around 60 households were surveyed twice, to capture information on the same household from both a male and female perspective. In almost 17 per cent of households, a discrepancy was recorded in the reported gender of the household head.

This chapter provides the results from these three forms of analysis of the survey data.

The focus group discussions allowed women, men, young women and young men to separately share their experiences through three participatory exercises. Exercises allowed for responses to open questions, which were compared to data from the defined questions of the quantitative survey. Local research teams in each country included male and female focus group facilitators and documenters, who led the focus groups with men and women respectively. The focus groups were conducted in the local languages of Pidgin in the Solomon Islands and Bislama in Vanuatu.

The first qualitative exercise was to construct a timeline of 'good and bad times' going back as far as ten years and as few as four. Participants started by listing good and bad times on pieces of card and would then arrange them into a timeline and discuss why events happened and what caused them. The second activity focused on the 'bad times' and ranked them according to their severity. The group then discussed what happened, who was affected, how they were affected and what people did to respond to their circumstances.

The last exercise of the focus group discussions focused on what could be done to improve circumstances – through a 'low hanging fruit' exercise. A tree would be drawn on a piece of butcher's paper or on the ground and participants would brainstorm what could be done and by whom. If things were easy to do – meaning able to be done within the village or by the family – they would be put in the low branches of the tree as 'low hanging fruit'. If other actions were harder or further away – if they required government action or seemed more difficult – they were placed in the higher branches of the tree.

After the focus group discussions, the facilitators of the various focus groups would come together and collate the main points of the discussions. Then, before leaving each community, the research teams provided feedback from the focus groups – particularly some of the ideas from the low hanging fruit exercise – at a community meeting.

The focus group exercises were initially analysed by the research teams in debrief sessions in Honiara and Port Vila. At these sessions, the research teams would combine the timeline, ranking and low hanging fruit exercises and analyse the trends and differences across them. The focus group discussions were translated from Pidgin or Bislama into English by members of the research team who speak and read these languages. After translation, the focus groups were analysed, systematically taking account of geographic, gender and age differences in responses. The timeline and ranking exercises were coded and a frequency analysis done by gender. The focus group data were also triangulated with information from the household surveys and key informant interviews.

3.3 Results: Gender-Differentiated Experiences of Economic and Other Shocks and Their Impacts

Through their discussions of good and bad times, as well as their ranking of such events, the focus groups identified the experience of various economic and other shocks. They also identified the impacts of such shocks on their own households as well as the community more broadly. In order to understand the focus groups' overall perceptions of the different impacts of shocks, a simple frequency analysis was applied to the events that focus groups, in aggregate, characterized as being 'bad' for their communities. Across focus groups in the Solomon Islands and Vanuatu, the most frequently reported 'bad' events were: (1) rising education costs; (2) rising food prices; (3) rising fuel prices and higher transport costs; (4) threats to social cohesion, such as violence, crime and substance abuse; and (5) unemployment. The gender differentiated experiences and responses to these main shocks and impacts are detailed in the sections below. The most relevant findings are highlighted, drawing on both the household survey and focus group discussions. This section of the chapter concludes by examining the sources of support sought by women and men during times of need.

3.3.1 Rising Education Costs

The household survey highlighted an underlying gender inequality in access to (particularly higher) education. There was only a slightly greater proportion of high school achievement – that is, having passed at least one year of secondary school – for men than women (36 per cent of adult men in the sample versus 31 per cent of adult women). However, the share of men that attended tertiary (or technical) schooling was double the share of women (10 per cent of adult males in the sample versus 5 per cent of women). While primary school is notionally free in both countries, it was found that rising costs of living were impinging upon the ability of households to manage the associated costs of their children's education. In particular, rises in the costs of transport (see below), clothing, food and books as well as fees (additional contributions asked by the school) were key concerns for both men and women in the Solomon Islands and Vanuatu. Across gender, it is expected that 'fathers and mothers must work hard' to generate income and manage school fees (Blacksands, Port Vila, Vanuatu Younger Men), and both genders highlighted the need to pay school fees in installments and offer payment in-kind, such as through volunteering on projects and working bees within the school to manage costs.

However, there is a more apparent 'gendered' distinction in regards to how women and men propose to seek external assistance in managing school fees. On the one hand, women focused on support from *wantoks* (a Pidgin English word meaning 'one who speaks the same language' – see Chapter 6), fundraising for school fees and drawing on family support in response to education costs (Malu'u, North Malaita, Solomon Islands Older Women; Pentecost, Vanuatu Younger

Women; Ohlen, Port Vila, Vanuatu Younger Women). On the other hand, men highlighted the responsibility of the government: 'Government must look at the issue of school fees' (Baravet, South Pentecost, Vanuatu Younger Men). 28 per cent of statements by men on education expected that the government provide free or subsidized education in comparison to 13 per cent of women's statements in the focus groups.

In focus group discussions, women's statements also more frequently identified quitting school as a response to the increased cost of education. There were reports of some girls quitting school in the Solomon Islands: 'Increased number of female dropouts because of school fee problems' (Maruiapa, East Guadalcanal, Solomon Islands Younger Women). Additionally, both younger women from Lilisiana and older women from Ambu – both in the town of Auki in the Solomon Islands – stated that after quitting school, a student can 'stay home'. While these two focus groups do not specifically mention the gender of the student, this statement could be suggestive of young girls' duties within the household as taking precedence over education.

3.3.2 Rising Food Prices

Findings from the household survey highlight that the experience of food price shocks was generally universal. That is, irrespective of the head of the households, or the gender of the respondent, a high proportion of households experienced rising food prices. Yet, on the face of it, it seems that women in general encountered greater difficulty coping with increased food prices than their male counterparts. Where both a woman and a man were surveyed in the same household, there were differences in the perception of the origin of food and food prices. Women, particularly in Vanuatu, were four times more likely than men to report that purchasing food had become increasingly difficult in the two years preceding the survey. They were also four times more likely than men to report that they had experienced a severe food price increase (that is, prices had 'gone up a lot' rather than had just 'gone up'). There were also differences in how women and men responded to questions about how they coped with rising food prices. Women and men indicated in approximately equal measure that adults went without food, although this survey question did not ask the gender of the adult that made this sacrifice. However, in other responses, it is evident that women disproportionately responded by having smaller meals and sharing food; whereas, men went to restaurants less and bought cheaper food.

Perhaps reflecting the relatively closer proximity of women to food preparation activities of the household, in both countries women were more likely than men to claim that the household was more frequently given food by relatives. In double-surveyed households, women were more likely than men to disclose that they had experienced different manifestations of food insecurity; this included being almost twice as likely to indicate that an adult had skipped meals in the year preceding the survey because there were insufficient funds to purchase food. Women were more likely to report that they had adjusted to food price inflation by purchasing cheaper and lower quality food and relied on gardens due to a lack of money.

The finding that women are disproportionately affected by higher food prices can be compared to the focus group discussions. There is particular reference to women's role in food provision: 'Women are most affected [by rising food prices] due to their responsibility to feed the family' (Ambu, Solomon Islands Older Women). Some women's focus groups linked rising food prices to health, highlighting that households were experiencing a nutritional deficit and unable to eat a balanced diet with the purchase of cheaper food. To manage the rising cost of food, both men and women identified the need to work more or obtain employment, and men and women equally noted the division of labour (that is, the 'paid jobs' that should be found were 'house girls [domestic jobs], labour jobs for men'). However, women's focus groups reported that they were also performing the 'roles of men' such as 'fishing, diving' (Lilisiana, Solomon Islands Older Women).

3.3.3 Rising Fuel Prices and the Cost of Transport

While there are geographic differences in how increased fuel and transport prices are experienced in the various surveyed communities, rising fuel prices were universally identified as difficult by focus groups. While rural communities experienced fuel price inflation through rising transport costs owing to their remoteness, urban communities, too, experienced rising fuel prices through a combination of rising prices for local transport as well as cooking fuel (rural communities generally rely on biofuels, such as firewood and coconut husks, for cooking). Similar to rising food prices, the experience of fuel price rises was generally the same between men and women respondents, yet women experienced greater negative impacts of this shock than men. Across all households, women respondents were almost twice as likely to report that it had become a lot harder to cope with rising fuel prices than men; a finding substantiated by data from within households when the duplicate surveys were considered.

There were also gendered differences in the way that women and men reported coping with fuel inflation. Male survey respondents were somewhat more likely than women to report that the household had reduced spending on cooking fuel and switched to a cheaper cooking fuel in response to a fuel price shock. However, in focus group discussions, women indicated that they responded to a fuel price shock by restricting their mobility. Women were much more likely to report travelling less or not at all to cope with increased transport costs, accounting for 20 per cent of their response statements to a fuel price shock, relative to just 2 per cent for men.

Women's focus group discussions also highlighted higher transport costs impacting on their children's education: This 'affects our children going to school – they miss classes when there is no bus fare and lunch money' (Lilisiana, Solomon Islands Older Women). Women also expressed concern that rising transport costs could negatively affect academic performance as children miss classes. In most cases, both women's and men's focus groups focused on ways to adapt to rising transport costs in order to keep children in school, while a few respondents did note that they had to change schools.

To manage rising fuel costs, both women's and men's focus groups highlighted working more and using alternative sources of fuel, including renewable energy and bio fuels. However, women tended to be more specific about local strategies and adaptations to offset the increased price of fuel, including the substitution of firewood or charcoal, the judicious use of cooking devices, and again reducing travel. On the other hand, men in Vanuatu saw the issue in terms of policy and highlighted the role of government in reducing fuel prices, in allocating resources and in controlling costs. Interestingly, this is an area with reference to global markets: 'Government must control the price of fuel as the world market price is more and more' (Mangalilu/Lelepa, Vanuatu Younger Men). Men in the Solomon Islands did not identify the government as responsible for addressing fuel prices; rather, they reinforced the women's strategies (travelling less and using more efficient power generation), but provided less detail about these coping strategies.

3.3.4 *Shocks and Social Cohesion*

There was a significant gendered difference in the identification and experience of 'social' impacts, such as violence, crime and substance abuse. Overall, women highlighted impacts at a personal, household and community level, noting the effect on children and that 'females are victims' (Malu'u, North Malaita, Solomon Islands Younger Women). Men tended to frame their observations only at the community level. For example, on account of criminal activity and a lack of respect and cooperation in the community, a third of men's responses from focus groups advocated for increased or renewed participation in traditions and attendance at church. A minority of men's responses (4 per cent) held gender equality laws responsible for the degeneration of community cohesion: 'Parliament must suspend laws about equal rights' and the community 'must go back to Biblical principles, which say that men are the head of the household' (Ohlen, Vanuatu Older Men). In response to religious or political disputes, a third of men's responses suggested engaging with government or 'us[ing] your democratic rights/voting rights wisely' in order to exert political pressure. In contrast, older women of Ambu, Solomon Islands noted that political instability did not tend to affect women because 'their voice has little or no place in most decision-making in the community'.

In particular, women were concerned with gender-based violence and home insecurity. One women's focus group noted that 'women are most vulnerable to violence and destruction' (Ambu, Solomon Islands Older Women). As mentioned previously, women were usually less likely to seek government assistance as a response to shocks; however, in discussions about issues of social cohesion, women's focus groups did advocate for the introduction of laws and punishment for crime, drug abuse and a lack of respect in their communities. In discussing violence and rape specifically, 50 per cent of women's responses held that the government was responsible for increasing security and that women had a role in 'electing good leaders into parliament' in order to ensure law and order. Men were silent on the topics of violence, rape and home security. Interestingly, with respect to crime, women's and men's focus groups tended to concur that increased job opportunities could reduce crime.

3.3.5 Employment and Other Work

The survey found that women are under-represented in formal employment and over-represented in the informal economy, most commonly through selling things at market (see Table 3.1). Across both of the countries surveyed, almost half of the jobs that were held by a man were in formal employment – that is, a government employee or a wage earner in a private enterprise. This compares with about 16 per cent of women's jobs. In contrast, more than 80 per cent of women's work was in the informal sector: peddling various items such as food (cooked and uncooked), jewellery, cigarettes and kava (in Vanuatu) or betel nut (in the Solomon Islands). In the instances where they had access to formal employment, women also generally reported lower income than men (58 per cent received less than SBD$650/VT7,500 per month (approximately US$75) and 78 per cent received less than SBD$1,300/VT15,000 (approximately US$150)). In contrast, 27 per cent of men's jobs earn less than SBD$650/VT7,500 per month and 50 per cent receive less than $1,300SBD/VT15,000 per month. Men reported working half a day more each week than women although this did not account for women's unpaid care work.

Table 3.1 Sources of income by gender in the Solomon Islands and Vanuatu (per cent)

Job Type	Men	Women
Formal employment	48.8	15.7
Informal markets	47.7	82.3
Other	3.5	2.0
Total	**100**	**100**

Note: Formal employment refers to a government employee or a wage earner in a private enterprise (including overseas seasonal worker schemes), while informal markets refers to peddling of various items. Includes only jobs that were identified as being held by either a man or a woman – excludes jobs that were identified as being held by both a man and a woman. Other includes remittances and gifts/patronage.

Source: The authors.

However, the definition of work in Melanesia is rather fluid. Waring and Sumeo (2010) note that a substantial amount of productive labour, of both men and women, is essentially 'invisible' in official labour and production statistics. Consequently, a substantial amount of productive, value-adding activities may be misclassified when labour is considered and organized according to employment status. This is evident in the primary activities of people who did not nominate that they had (formal or informal) employment (see Figure 3.1). Reflecting their traditional role as caregivers and custodians of the home, a high proportion (over 50 per cent) of women reported that they did housework and cared for children – a much higher proportion than men. Men, in contrast, were more likely to engage in construction activity and hunting than women – again reflecting the traditional gendered division of labour. Responsibilities in the garden were much more evenly shared. Interestingly, when specifically asked about which gender is primarily responsible for the garden, rural communities were more likely than urban communities to indicate that responsibility was explicitly shared (67 per cent compared with 56 per cent) potentially due to the greater need for formal employment in urban areas, reducing the number of people available to tend to the garden.

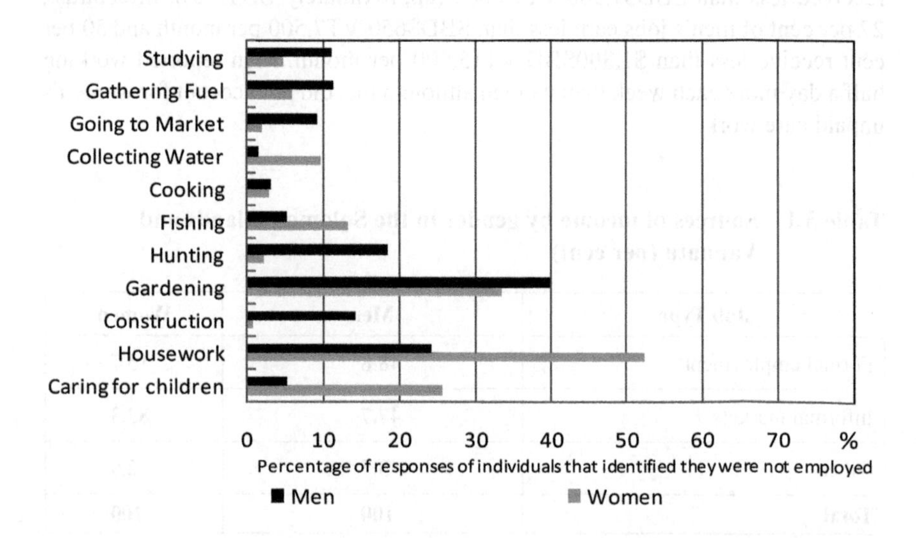

Figure 3.1 Activities of men and women who do not work in the Solomon Islands and Vanuatu (per cent)

Source: The authors.

Both men and women identified similar strategies to respond to falls in real household income, either by a loss of employment or by a sudden rise in prices, caused by economic shocks. In both surveys and focus group discussions, it was evident that women and men had sought to augment household income to make ends meet. When asked about their primary strategies for earning income, responses reflected the traditional, and emerging, gendered divisions of labour. Women were more likely than men to identify increased participation in existing labour market systems, including increased peddling of garden produce or other items (such as copra, cocoa, handicrafts, fish, ice blocks and sewing). Yet, they were also more likely than men to suggest looking for additional formal employment. Men, in contrast, were more likely to nominate starting a new business (such as construction, driving a bus/taxi, opening a *nakamal* (in Vanuatu) and selling betel nut (in Solomon Islands)) – though formal employment was also nominated.[2]

However, while women and men both attempted to increase their contribution to household income through increased labour output, women were still expected to maintain their role and their responsibilities as a caregiver around the home. This is consistent with Waring and Sumeo's (2010) finding that women have multiple roles in sustaining the household and face a great burden on their time.

There were also considerable differences in the way that women and men identified how they could foster resilience to falling real household incomes in the future. In focus groups, women highlighted the importance of budgeting and the employment of more household members because 'the money which the one person earns is not enough to feed the family' (Pentecost, Vanuatu Older Women). In the Guadalcanal Plains Palm Oil Limited (GPPOL) communities of the Solomon Islands, where rising commodity prices contributed to windfall gains to out-growers of palm oil, women noted that effective budgeting and savings were needed in order for households to be financially independent in both upswings and downswings in commodity prices and 'pave way for future investment in self-managed businesses'. As in other areas, men in almost a quarter of responses advocated that the government should act to increase their income (through higher minimum wages or facilitating access to overseas markets) or facilitate their businesses (by reducing tax or providing services, subsidies and support). Women, again, did not address the role of government.

3.3.6 Sources of Support

Women and men drew on a different mix of social support to deal with negative shocks (see Figure 3.2). Across households, women tended to draw relatively more on friends and neighbours and the church, while men drew relatively more on family. The double-surveyed households, however, indicated that, within households, women were only marginally more likely than men to draw on such

2 Interestingly, roughly equal shares of women and men (one in six), volunteered that they did not know how their household would generate additional income.

external assistance. Interestingly, men were more likely than women to jettison social commitments, being more likely to reduce their contributions to *wantok* and fundraising (though not the church). That men were more inclined than women to reduce social contributions was confirmed in the double-surveyed households. In the focus groups, men frequently highlighted the role of the government in responding to the impacts of increased prices and other 'bad times'. However, women's focus groups rarely discussed the role of government and were more likely to consider what they could do for the family or in the community.

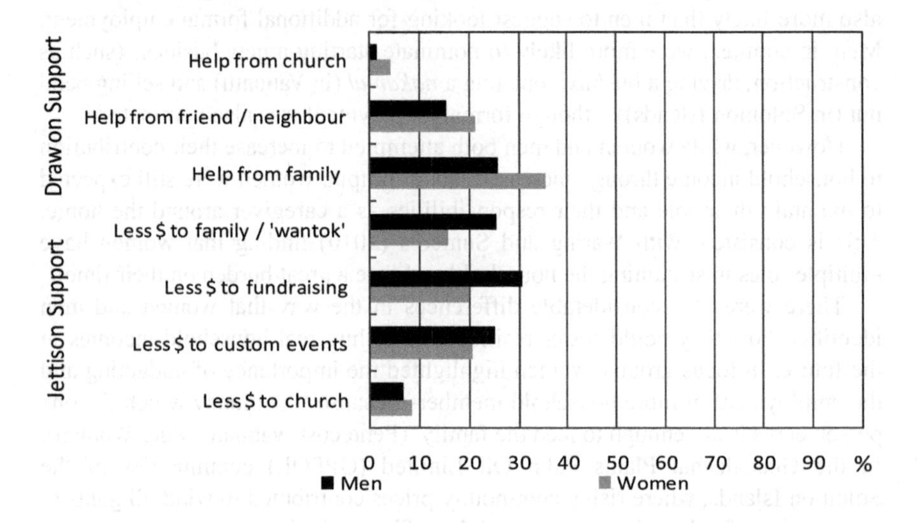

Figure 3.2 Responses to a negative economic shock by gender in the Solomon Islands and Vanuatu (per cent)

Source: The authors.

3.4 Discussion: Gender Roles and Inequalities in the Solomon Islands and Vanuatu and International Evidence of Impacts from Shocks

Women in the Solomon Islands and Vanuatu have borne a significant burden in the adjustment to the recent food, fuel and economic crises in the world economy. Women have faced the challenge of holding household finances, assets and food supplies together as real incomes fall by seeking additional sources of money and support as well as being parsimonious with expenditures on food and fuel. Concomitantly, they have maintained their roles as caregivers and food providers. They have been exposed to violence and heightened levels of stress. There were also some key differences in the ways that women and men articulated their responses to such shocks. Women focused more on immediate solutions, including drawing more heavily on friends, neighbours and other local networks and highlighted the role of these relationships in their suggestions for future improvements. Men, in contrast, concentrated on institutional responses, advocating for both a strengthening of tradition and increased support from government.

Both the quantitative and qualitative data from this research indicate that the experience of, and responses to, recent economic shocks are reflective of distinctions, which are linked to gender roles and inequalities in the Solomon Islands and Vanuatu. Over time, women's and men's roles, responsibilities and expectations in these countries have developed and have been influenced by *kastom* or tradition, colonialism, churches, modernization and globalized concepts of gender equality and human rights (as discussed below with reference to Jolly, 1991, 1994, 1996; Sillitoe, 2000; Bolton, 2003; Bowman et al., 2009; Wallace, 2011). This section of the chapter reviews the research findings outlined above in light of analysis contained in the existing literature on women's difficulties in income generation, responsibilities for unpaid work, sources of support and experiences of violence specific to the Solomon Islands, Vanuatu and the Pacific region as well as the gendered dimensions of recent crises.

3.4.1 Difficulties in Income Generation

While both men and women recognized the necessity of generating income as a response to price shocks, women were more apt to increase participation in informal markets or to depend on better budgeting to increase cash. As the results from this study indicate, over 80 per cent of women reported working in the informal sector while men were more equally represented in both the formal and informal sectors. In developing countries, 60 to 90 per cent of the workforce is located in the informal economy (ILO, 2002 as cited in King and Sweetman, 2010), and over half of women worldwide are considered to be in 'vulnerable jobs' because they work in the informal economy through self-employed means or as unpaid family workers (Seguino, 2010, p. 185).

However, informal economy impacts were often unrecorded in official statistics or monitoring of the GEC (Green et al., 2010). From an examination of the informal sectors in Asia, Latin America and sub-Saharan Africa, the recent economic crisis has increased competition for jobs and wages in the informal sector, especially as the recently retrenched and underemployed enter the informal sector, impacting on women's income and indicating that 'the informal economy does not serve as a safety net' (Horn, 2010, p. 264). Workers in the informal sector as well as the self-employed are offered social protections in only one third of developing countries (ILO, 2009 as cited in King and Sweetman, 2010). Such observations underline the limitations of household resilience and dependence on the informal sector in the absence of social protections. Solomon Islander and ni-Vanuatu women make significant efforts to maintain the well-being of their families during a crisis. However, their actions can only go so far without broader support. Further impediments to women's employment include discrimination and unclear legal frameworks (Hedditch and Manuel, 2010a, 2010b) and intimidation and sexual harassment (Maebuta and Maebuta, 2009), while a lack of access to training, transport, infrastructure and assistance impacts on women's ability to most productively use gardens, markets, skills and time (Jansen et al., 2006; Vunisea, 2007).

Women's education is an essential factor in achieving gender equality. Women's equality is synonymous with 'rights, resources and voice' (Morrison et al., 2007, p. 2). This improves women's education, health and market access, which in turn increases women's opportunities in the labour force and income generation (Morrison et al., 2007). Such opportunities contribute to women's increased influence over expenditure and savings as well as household decision-making (Morrison et al., 2007). Better education and health and improved decision-making – and by implication, increased labour force participation – in turn benefit children's welfare, health and education, potentially decreasing current and future poverty and inciting economic growth (Morrison et al., 2007). Beyond these instrumental reasons, there is the more fundamental intrinsic value of access to education being offered equally to boys and girls. While Vanuatu has increased gender parity in primary school education rates, women and girls remain under-represented at the vocational, technical and tertiary levels (Strachan, 2004; Bowman et al., 2009); are less likely to apply for scholarships (Strachan, 2002 as cited in Strachan, 2004; Bowman et al., 2009); and may not have the educational background to apply for certain jobs (Vanuatu Statistics Office, 2000 as cited in Strachan, 2004). In the current economic climate, there is some evidence from this study and the existing literature that food price shocks have strained access to education. In Vanuatu, enrollment dropped by 20 per cent at one secondary school, and rice shortages have forced temporary closures at other schools (Miskelly et al., 2011). The impact of ongoing, multiple crises can be significant: research with informal workers in Africa, Asia and Latin America in 2009 and 2010 found a higher rate of school withdrawals as the crises continued to affect informal workers (Horn, 2011). During the recent crises and past crises, other studies have found that impacts on education are mixed and can include the withdrawal of children from school (Parks et al., 2009; UNICEF, 2009; Patel, 2009; World Bank, 2008), or, in some contexts, a prioritization of spending on children's education (Green et al., 2010; Hossain and Green, 2011).

In line with other research (Bowman et al., 2009), this study has found that concerns about children and the cost of their education as well as the household continue to drive women into the cash economy. In Papua New Guinea's mining industry, a group of educated and employed women, regardless of their marital status, prioritized education by paying schools fees for at least two children, either their own or within their extended families, on account of its promise of social mobility (Macintyre, 2006). For a similar reason, women's focus group discussions in the Solomon Islands and Vanuatu outlined particular anxiety around the impact of increased fuel and transport costs on children's education because 'education is the only way to better life' (Ambu, Solomon Islands Younger Women). While

Melanesian women contribute to household income and participate in the cash economy in order to pay for school fees and other necessities, data from this research indicate that this can prove difficult on account of women's under-representation in the formal sector, generally lower incomes and pressures to balance income-earning activities with housework and care work.

Perhaps, these difficulties in income-earning activities may be understood through observations of the value placed on women's and men's work in Melanesia. In South Pentecost, Vanuatu, a gendered division of work has been observed for tasks, such as textile production, the planting and preparation of crops (such as yams), fishing and pig herding (women most often attend to the daily feeding and care of pigs which are owned only by men) (Jolly, 1994). Once such goods are brought to public exchange, Margaret Jolly (1994, p. 85) offers, 'Although conjoint labour is involved in producing both yams and pigs, male and female labour is not accorded equal value. Female labour though acknowledged is ultimately of lesser value, in the representing of these goods as 'male'…in the context of exchange'. Other responsibilities, such as cooking, gardening, processing copra and caring for children are not rigid in terms of division of work; however, men generally contribute less to responsibilities associated with childcare (Bolton, 2003). Responsibility for children along with women's other unpaid work will be further explored in the following sub-section.

3.4.2 *Responsibility for Unpaid Work*

As communicated in the results of this research, women in the Solomon Islands and Vanuatu continue their unpaid responsibilities of housework, care work and food provision alongside trying to generate income in the informal or formal economies. Responsibility for household members and care work 'falls within the framework of relationship nurturing, reciprocity and service to the family and community' (Waring and Sumeo, 2010, p. 4). On account of the number of unpaid responsibilities undertaken by women in the Pacific region, women are considered time poor, overworked and ill-compensated (Waring and Sumeo, 2010). Analysing the extent and impact of unpaid care work on individuals' time in the Pacific region is made difficult by a lack of recent time use studies, which also limits the ability of policymakers to address the issue through targeted policy (Waring and Sumeo, 2010). Vunisea (2007, p. 26) adds that '[t]he continuing classification of women as food foragers for family consumption…translates to the continuing neglect of women in mainstream development initiatives, education and training' because the work involved in gathering, preparing and selling food is labeled as a traditional, household responsibility as opposed to a formal, economic activity.

Women's focus group statements indicate that women may uniquely respond to and experience the pressures of cost increases. This was especially highlighted in discussions around food and fuel prices as women spoke of their worry and stress about the impacts on their children and households. Women's focus groups reported that they were concerned about running out of food and, therefore, had smaller meals themselves and shared food. This is consistent with evidence in the international literature that women often cut back on their own consumption to allow greater consumption by men or children as the quality and quantity of food degrade (Green et al., 2010; King and Sweetman, 2010). Many such 'coping strategies' cannot be maintained and, instead, can be seen as 'desperation measures' (Green et al., 2010, p. 27, 36; King and Sweetman, 2010, p. 11). However, women often forego food even in times of non-crisis. For instance, prior to the food crisis, women went without meals more than men in nearly 60 per cent of households in Bangladesh (Quisumbing et al., 2008). This is particularly concerning since malnutrition and an unbalanced diet can severely impact women's health, especially during pregnancy, and cause micronutrient deficiencies (Holmes et al., 2009).

In this study, women in the Solomon Islands and Vanuatu were more expressly concerned with the impact of decreased quality and quantity of food on children's health as opposed to on their own health. At the extreme, economic shocks can increase child mortality through reductions in food and nutrition as well as decreases in social spending resulting in fewer public services (Parks et al., 2009; Patel, 2009; Harper et al., 2009). The reduction in food and other coping strategies can have long-lasting effects on the health and opportunities of the individual (World Bank, 2008; Horn, 2009; Mendoza, 2009) as well as more macroeconomic effects by potentially impeding the future labour force and economic growth (World Bank, 2008; Mendoza, 2009; UNICEF, 2009; King and Sweetman, 2010). During crisis, pressures to provide for the family and a lack of affordable public services overtaxes parental time, reducing their availability to provide nurture and care for their children (Harper et al., 2009). Women are balancing the physical and emotional needs of the household, sometimes at the expense of their own. In Nigeria, for example, demands on women in the wake of the food and fuel crisis gave rise to feelings of anxiety and depression in some women (Samuels et al., 2011).

As a response to increased fuel costs, women in this study again changed their own behaviour so as to contribute to household resilience. Women reported limiting their travel while men did not, perhaps suggesting a division of labour that women focus on the home, garden and local market and men often travel further for work. In Vanuatu, men's migration to other islands or overseas for wage labour impacted women: 'With many young men away, more work devolves on the women, children and older men in the community. Women whose husbands are away working are especially badly off, since they often have an enormous burden with garden work, gathering, child care, etc.' (Jolly, 1994, p. 88–9). During the New Zealand seasonal employment programme, women's participation was dependent upon community decision-making, and some communities denied women's participation in the programme, in spite of the personal and economic benefits, because it was thought to be detrimental to community structures and traditional gender roles (Bowman et al., 2009). Additionally, for cooking purposes, this research found that women in the Solomon Islands and Vanuatu have substituted charcoal and firewood for fuel.

However, collection of firewood is another demand on women's time, contributing to women being time poor (Mendoza, 2009; Samuels et al., 2011).

3.4.3 *Sources of Support and Institutional Structures*

Overall, women in this study tended to associate far more household impacts with increased prices and social issues than men did. Women framed their responses to shocks around concern for children and family members and sought to take corrective action themselves – making their responses personal, tangible and local. Men, on the other hand, more often discussed responses to shocks in terms of institutional action – by the government, by the church and in line with community tradition.

Men in one focus group discussed how to address the social impacts of economic shocks with reference to men's and women's status. As noted in sub-section 3.3.4, this focus group called for an abandonment of equal rights legislation and a reinvigoration of 'Biblical principles' to reinstate men as the household head (Ohlen, Vanuatu Older Men), and other male focus groups discussed increased church attendance as a response. In other research, it was found that some men from the Solomon Islands, Vanuatu and Papua New Guinea have cited and interpreted Bible passages as support for male dominance and physical discipline of women (Jolly, 1996). Churches, established by missionaries, did not allow the ordination of women or for women's leadership in positions above men (Bolton, 2003). Even though men compose the main church structures and hierarchy, church groups have provided fora for women to support one another (Wallace, 2011). However, some married men in the western Solomon Islands have dissuaded their wives from participating in church activities, claiming that women would be abandoning their homes, children and husbands by participating (Jolly, 1996).

In the South of Vanuatu during the period 1848–70, Presbyterian missions introduced the concept of 'wifely domesticity' (Jolly, 1991, p. 32). These missionaries saw women's manual work outside the home as contrary to their Christian ideals, and while their 'project of domestication' brought about some improvements for women, it 'also imparted a new model of male domination, predicated on devaluation of women as domestic beings and a celebration of men as public beings' (Jolly, 1991, p. 45). Sillitoe (2000) puts forth that the household roles and activities of Melanesian women and men have distinct domains, the domestic and public respectively, and these activities complement and do not subjugate or favour one over the other; rather, 'it is the so-called modernization process itself that introduces the notion of a sexual status hierarchy', with missionary efforts often initiating this process (Sillitoe, 2000, p. 102). In light of observations about church structures and missionary activities, men may be prone to seek assistance and support from institutions which speak to male dominance.

Women, on the other hand, may be less inclined to solicit assistance from some institutions on account of a lack of representation. In this study, women's focus groups rarely identified government assistance as a mechanism for addressing shocks and their impacts. Across Pacific Island countries, women have only about 5 per cent representation in parliament, with one Solomon Islander woman and no ni-Vanuatu women currently serving in their respective national Parliaments (Pacific Women in Politics, n.d.). Additionally, the governments of the Solomon Islands and Vanuatu have not comprehensively addressed gender inequality (Wallace, 2000; Wallace, 2011). While non-government organizations have worked to address gender issues in these countries, cooperative efforts between the government and non-government sectors are needed to address gender issues in the Solomon Islands and Vanuatu (Wallace, 2000; Wallace, 2011).

As discussed by a women's focus group in Ambu, Solomon Islands, Melanesian women are seldom involved in decision-making processes. For instance, 69 per cent of women in Vanuatu and 58 per cent of women in the Solomon Islands in a partner relationship reported that 'they have experienced some sort of controlling behaviour by their partners' in regards to their decision-making, accessing healthcare and regulating mobility and family visits (World Bank, 2012, p. 99). Wallace (2011) adds that rural women, in particular, are often disadvantaged by an unawareness of their own rights and by exclusion from formal decision-making, and that 'entrenched traditional views' pose a challenge to addressing gender inequality in these countries (p. 509). In some contexts, the household often spends less on nutrition, health and education when women have less input into financial decision-making processes (Holmes et al., 2009). Such exclusion of women from decision-making has meaningful consequences for the female individual and the household in times of crisis and non-crisis.

3.4.4 *Experiences of Violence*

Incidents of violence were raised predominantly by women's focus groups as issues in their communities. This information cannot be interpreted as suggesting that violence against women and women's feelings of insecurity happened as a direct result of economic shocks. At this point in time, it would also be unfair to suggest a correlation between this data and economic shocks. Women focus-group participants in the Solomon Islands and Vanuatu regard violence and insecurity as important issues to be addressed, and discussion on these topics is suggestive of vulnerability to violence, irrespective of a link to economic shocks.

The literature indicates that violence against women, as well as violence against children, often increase in times of economic hardship (Harper et al., 2009; Parks et al., 2009). In some developing countries, qualitative data suggests that the stress of increased unemployment and decreased incomes as a result of the financial, food and fuel crises was found to contribute to arguments at home, violence against women by men and violence against children by women (Heltberg et al., 2012). Phua (2011, p. 2) offers: 'The financial and psychological stresses (including self-esteem problems) arising from prolonged unemployment may increase the likelihood of domestic violence within affected families'. Indeed, it has been suggested that the stress of rising food prices in Bangladesh, Indonesia, Kenya and Zambia impacted men's sense of self-worth in providing for their families and resulted in substance abuse, household arguments and domestic violence (Hossain and Green, 2011).

Rates of gender-based violence are high in the Solomon Islands and Vanuatu. Many women suffer physical or sexual violence at the hands of an intimate partner, with 64 per cent of Solomon Islander women, aged 15 to 49 (Secretariat of the Pacific Community, 2009), and 60 per cent of ni-Vanuatu women reporting such experiences of intimate partner violence (Vanuatu Women's Centre, 2011). Barriers to reducing violence against women in Melanesia include women's low status and economic dependence, and attitudes and some customary practices (Ellsberg et al., 2008).

Practices, such as bride price, arranged marriages and forced marriages, often upheld under customary law, are deleterious to women and women's equality in the Pacific Islands (Jalal, 2009). Additionally, most Pacific Island countries apply customary law 'either by social sanction or by the conventional courts' (Jalal, 2009, p. 8), leaving women few channels formally and legally to address practices that perpetuate gender inequality and violence (Jalal, 2009). In rural Vanuatu, where the majority of the country's population lives, custom law as opposed to the formal legal system is often applied by, usually male, chiefs (Bowman et al., 2009). Moreover, the police force in Vanuatu lacked a formal response procedure to violence against women (AusAID, 2009), and rape and domestic violence are underreported to the police (Bowman et al., 2009). In the Solomon Islands, it was found that women have less confidence than men in the police and justice system and are less likely to report crime than men are (AusAID, 2009).

Despite government policies to address gender issues in the Solomon Islands and Vanuatu, the benefits and accompanying changes in attitude towards women are yet to become apparent (Wallace, 2011). In this study, men's focus groups called for the suspension of equal rights legislation and renewal of tradition, while women's focus groups acknowledged their limited influence over decision-making and susceptibility to violence. Jolly (1996), noting the use of universal human rights language in discussions about domestic violence and measures to address it in Vanuatu, interprets that '[i]n acting against the power of western-derived laws and the expatriate male judiciary, some men at both local and national levels are not so much reclaiming powers which they had in *kastom* as asserting new and more strenuous forms of male control over women, in contestation with outside powers and foreign values' (p. 182). It is recognized that women and men must create *kastom* together, with the understanding that human rights do not have to be incompatible with *kastom* (Jolly, 1996).

As previously discussed, female political representation is extremely low in Pacific Island countries, ranking among the worst globally (Hedditch and Manuel, 2010a, 2010b; World Bank, 2012). This impairs advocacy for gender equality as 'women's chronic under-representation in economic, political and legal institutions across the [Asia-Pacific] region has produced deficits in power and voice, which in turn allow inequalities to go unchallenged' (UNDP, 2010, p. 24–5). The focus group statements on issues of violence and gender dynamics may contribute to an understanding of how community, *kastom*, law and female empowerment are keys to maximizing resilience.

3.5 Conclusion

Women sit at a peculiar juncture in the Melanesian societies of the Solomon Islands and Vanuatu. They experience inequality in their households, workplaces, communities, societies and institutions. However, they are economically vital to the household. Using both quantitative and qualitative data from urban and rural communities, this research finds that the impacts of economic shocks elicited particular responses from women in the Solomon Islands and Vanuatu, and women's responses were often shaped by and reinforced gender roles and inequalities in these countries. In the event of an economic shock to a household, Solomon Islander and ni-Vanuatu women communicated their attempts to absorb the household's hardship. They continued care work, housework and food provision for the household and increased their participation in the cash economy. Where falls in household income could not be mitigated, there is evidence from this research that women tried to offset rising food and fuel costs by consuming or purchasing less food, traveling less or looking to alternative fuel sources. The social consequences of shocks included the inability to pay for fees and other education-related expenses. Feelings and experiences of stress, insecurity and violence were also reflected in this data.

Policy considerations should include ensuring and improving women's access to education and training and the provision of assistance to the informal sector. These measures may bolster women's decision-making capacity and participation in the cash economy, which could benefit the household in turn. The formalization of procedures regarding violence against women and access to legal services is paramount to women's safety, and women must be made aware of the resources available to them so as to understand their options if they encounter violence or disempowering treatment. International pressures and support can also assist in bringing issues of gender inequality to the forefront.

While women's rights and gender equity remain important to social justice and to local, national and international development, the current marginalization of women's voices at the household and national levels indicate that gender inequality in the Solomon Islands and Vanuatu may only erode slowly. While more is to be understood about the impact of shocks on gender in Melanesia, this data identifies the unique situation of women in the Solomon Islands and Vanuatu in how they interact with and are represented in their local and national economies, how they influence shock management and how they attempt to minimize vulnerability and maximize resilience within their households in responding to economic shocks.

References

AusAID (2009), Responding to Violence Against Women in Melanesia and East Timor: Australia's response to the ODE report (Australian Agency for International Development: Canberra).

Bolton, L. (2003), *Unfolding the Moon: Enacting Women's Kastom in Vanuatu* (University of Hawai'i Press: Honolulu).

Bowman, C., Cutura, J., Ellis, A. and Manuel, C. (2009), *Women in Vanuatu: Analyzing Challenges to Economic Participation* (The International Bank for Reconstruction and Development/The World Bank: Washington).

Ellsberg, M., Bradley, C., Egan, A., Haddad, A. (2008), Violence against Women in Melanesia and East Timor: Building on Global and Regional Promising Approaches (Office of Development Effectiveness, Australian Agency for International Development: Canberra).

Green, D., King, R. and Miller-Dawkins, M. (2010), The Global Economic Crisis and Developing Countries, Oxfam Research Report, (Oxfam International: UK).

Harper, C., Jones, N., McKay, A., and Espey, J. (2009), Children in Times of Economic Crisis: Past lessons, Future Policies, Background Note (Overseas Development Institute: London).

Hedditch, S. and Manuel, C. (2010a), Solomon Islands: Gender and Investment Climate Reform Assessment, in Partnership with AusAID (International Finance Corporation: Washington).

Hedditch, S. and Manuel, C. (2010b), Vanuatu: Gender and Investment Climate Reform Assessment, in Partnership with AusAID (International Finance Corporation: Washington).

Heltberg, R., Hossain, N., Reva, A., and Turk, C. (2012), Anatomy of Coping: Evidence from People Living through the Crises of 2008–11, Policy Research Working Paper 5957 (World Bank, Social Development Department: Washington).

Holmes, R., Jones, N. and Marsden, H. (2009), Gender Vulnerabilities, Food Price Shocks and Social Protection Responses, Background Note (Overseas Development Institute: London).

Horn, Z.E. (2009), No Cushion to Fall Back On: The Global Economic Crisis and Informal Workers, Inclusive Cities Study, [accessed March 2013] http://www.inclusivecities.org/research/informal-economy-monitoring-study/.

Horn, Z.E. (2010), The Effects of the Global Economic Crisis on Women in the Informal Economy: Research Findings from WIEGO and the Inclusive Cities Partners, *Gender and Development*, 18(2): 263–76.

Horn, Z.E. (2011), Coping with Crises: Lingering Recession, Rising Inflation, and the Informal Workforce, Inclusive Cities Study, [accessed March 2013] http://www.inclusivecities.org/research/informal-economy-monitoring-study/.

Hossain, N. and Green, D. (2011), Living on a Spike: How is the 2011 Food Price Crisis Affecting Poor People? Oxfam Research Report, (Oxfam Great Britain: UK).

Jalal, I. (2009), Harmful Practices Against Women in Pacific Island Countries: Customary and Conventional Laws (United Nations Division for the Advancement of Women, United Nations Economic Commission for Africa: Expert Group Meeting on good practices in legislation to address harmful practices against women: Addis Ababa).

Jansen, T., Mullen, B.F., Pollard, A.A., Maemouri, R.K., Watoto, C. and Iramu, E. (2006), Solomon Islands Smallholder Agriculture Study: Volume 2 Subsistence Production, Livestock and Social Analysis (Australian Agency for International Development: Canberra).

Jolly, M. (1991), 'To Save the Girls for Brighter and Better Lives': Presbyterian Missions and Women in the South of Vanuatu: 1848–1870, *The Journal of Pacific History*, 26(1): 27–48.

Jolly, M. (1994), *Women of the Place: Kastom, Colonialism and Gender in Vanuatu* (Harwood Academic Publishers: Chur, Switzerland).

Jolly, M. (1996), 'Woman Ikat Raet Long Human Raet O No?': Women's Rights, Human Rights and Domestic Violence in Vanuatu, in A. Curthoys, H. Irving and Martin, J. (eds), *The World Upside Down: Feminisms in the Antipodes. Feminist Review*, 52: 169–90.

King, R. and Sweetman, C. (2010), Gender Perspectives on the Global Economic Crisis, Oxfam International Discussion Paper, (Oxfam International: UK).

Macintyre, M. (2006), Women Working in the Mining Industry in Papua New Guinea: A Case Study from Lihir, in K. Lahiri-Dutt, and M. Macintyre (eds), *Women Miners in Developing Countries: Pit Women and Others* (Ashgate: UK).

Maebuta, H.E. and Maebuta, J. (2009), Generating Livelihoods: A Study of Urban Squatter Settlements in Solomon Islands, *Pacific Economic Bulletin*, 24(3): 118–31.

Mendoza, R.U. (2009), Aggregate Shocks, Poor Households and Children: Transmission Channels and Policy Responses, *Global Social Policy*, 9(1): 55–78.

Miskelly, R., Cocco-Klein, S. and Abbott, D. (2011), Situation Monitoring: Food Price Increases in the Pacific Islands, UNICEF Working Paper (UNICEF Pacific: Suva).

Monk, J. and Hanson, S. (1982), On Not Excluding Half of the Human in Human Geography, *The Professional Geographer*, 34(1): 11–23.

Morrison, A., Raju, D. and Sinha, N. (2007), Gender Equality, Poverty and Economic Growth, Policy Research Working Paper 4349 (The World Bank, Gender and Development Group, Poverty Reduction and Economic Management Network: Washington).

Otobe, N. (2011), Global Economic Crisis, Gender and Employment: The Impact and Policy Response, Employment Sector, Employment Working Paper No. 74 (International Labour Organization: Geneva).

Pacific Women in Politics (n.d.), National Women MPs, [accessed 25 March 2013] http://www.pacwip.org/women-mps/national-women-mps/.

Parks, W., with Abbott, D. and Wilkinson, A. (2009), Protecting Pacific Island Children and Women During Economic and Food Crises: A Working Document for Advocacy, Debate and Guidance (UNICEF Pacific, UNDP Pacific Centre, UNESCAP Pacific Operations Centre: Suva).

Patel, M. (2009), Economic Crisis and Children: An Overview for East Asia and the Pacific, *Global Social Policy*, 9(1): 33–54.

Phua, K-L. (2011), Can We Learn From History? Policy Responses and Strategies to Meet Health Care Needs in Times of Severe Economic Crisis, *The Open Public Health Journal*, 4: 1–5.

Quisumbing, A., Meinzen-Dick, R. and Bassett, L. with Usnick, M., Pandolfelli, L., Morden, C., and Alderman, H. (2008), Helping Women Respond to the Global Food Price Crisis, Policy Brief No.7 (International Food Policy Research Institute: Washington).

Samuels, F., Gavrilovic, M., Harper, C. and Niño-Zarazúa, M. (2011), Food, Finance and Fuel: the Impacts of the Triple F Crisis in Nigeria, with a Particular Focus on Women and Children, Background Note (Overseas Development Institute: London).

Secretariat of the Pacific Community for Ministry of Women, Youth and Children's Affairs and the National Statistics Office (2009), Solomon Islands Family Health and Safety Study: A Study on Violence Against Women and Children, (Secretariat of the Pacific Community, Noumea).

Seguino, S. (2010), The Global Economic Crisis, its Gender and Ethnic Implications and Policy Responses, *Gender and Development*, 18(2): 179–99.

Sillitoe, P. (2000), *Social Change in Melanesia: Development and History* (Cambridge University Press: Cambridge).

Silva, J.A. (2009), Nicaragua: Women Bear the Brunt of Crisis, Inter Press Service News Agency [accessed March 2013] http://www.ipsnews.net/2009/06/nicaragua-women-bear-the-brunt-of-the-crisis/

Strachan, J. (2004), Gender and the Formal Education Sector in Vanuatu, *Development Bulletin,* no. 64: 73–7.

UNICEF (2009), A Matter of Magnitude: The Impact of the Economic Crisis on Women and Children in South Asia (UNICEF ROSA: Nepal).

UNDP (2010), Asia-Pacific Human Development Report: Power, Voice and Rights: A Turning Point for Gender Equality in Asia and the Pacific (Macmillan Publishers India Ltd. for the United Nations Development Program: India).

Vanuatu Women's Centre in partnership with the Vanuatu National Statistics Office (2011), Vanuatu National Survey on Women's Lives and Family Relationships, (Vanuatu Women's Centre: Port Vila).

Vunisea, A. (2007), Women's Changing Participation in the Fisheries Sector in Pacific Island Countries, *SPC Women in Fisheries Information Bulletin,* no. 16: 24–6.
Wallace, H. (2000), Gender and the Reform Process in Vanuatu and Solomon Islands, *Development Bulletin,* no. 51: 23–5.
Wallace, H. (2011), Paddling the Canoe on One Side: Women in Decision-Making in Vanuatu and the Solomon Islands, *Development,* 54(4): 505–13.
Waring, M. and Sumeo, K. (2010). Economic Crisis and Unpaid Care Work in the Pacific. United Nations Development Programme (UNDP). [Accessed March 2013] Retrieved from: http://marilynwaring.com/unpaidcarework.pdf.
World Bank (2008), Rising Food and Fuel Prices: Addressing the Risks to Future Generations, (Human Development Network and the Poverty Reduction and Economic Management Network, World Bank: Washington).
World Bank (2012), Toward Gender Equality in East Asia and the Pacific: A Companion to the World Development Report – Conference Edition, (World Bank: Washington) License: Creative Commons Attribution CC BY 3.0.

Chapter 4
Mobility and Economic Resilience in Melanesia

Alberto Posso and Matthew Clarke

4.1 Introduction

Migration is a normal and common human occurrence. Moving to seek new opportunities, new lands, new freedoms, fleeing persecution or economic stagnation is a phenomenon that has shaped and continues to shape human societies across the world. Migration from rural to urban centres is certainly a feature of the modern nation-state, with economic and other shocks sometimes playing an important role in the decision to move. Household shocks are an increasingly important feature of Melanesian life. They include idiosyncratic shocks (specific to the household) such as the loss of a garden to flooding or the death or illness of a household member as well as covariate (or community wide) shocks such as large natural disasters and price hikes of commodities that households have become dependent upon

Much migration is now urban migration. Migrants to urban areas can gain employment allowing them to remit money back to their families in rural areas. Employment opportunities though can be limited resulting in migrants simply moving from rural to urban poverty or hardship. Family reasons, the loss of a job or a shortage of money can lead to return migration, with people living in towns and cities deciding to return to the rural areas in which they had previously lived.

Within the Pacific, and Melanesia in particular, the geographic diversity of small sparsely populated islands has resulted in a long history of migration across waters (Lee, 2009). Both short-term and longer-term (or permanent) migration have been vital factors in economic development through local trade and accessing (since colonial times) a range of introduced welfare services, such as education and health care. A consequence of migration within Melanesia has been the increasing urbanization of a small number of centres throughout this region. Connell (2011) suggests that when Papua New Guinea is excluded, half of all Pacific Islanders now live in urban centres. However, in the case of Melanesia, urbanization has not been as strong. For example, the population of Honiara and Port Vila (the capital cities of Solomon Islands and Vanuatu respectively) are estimated to be 79,000 and 40,000, compared to their populations of 538,148 and 239,651 respectively (World Bank, 2012).

This chapter examines the extent and impact of migration in Pacific countries. It also investigates some of the consequences of urbanization and the role of internal remittances in providing economic resilience. It focuses on the Solomon Islands and Vanuatu using a rich new dataset from a survey of over 1,000 households as well as drawing on numerous focus group discussions and key informant interviews. In an attempt to capture the diversity in experiences of vulnerability and resilience, six locations were targeted in each country (two urban and four rural locations). It focuses on the impacts and responses of households to the recent hikes in the prices of both food and fuel and to the Global Economic Crisis (GEC). Household surveys in the Solomon Islands were conducted in Honiara (the capital), Guadalcanal Plains Palm Oil Limited (GPPOL) villages and the Weather coast on the main island of Guadalcanal, Auki (the second largest town) and Malu'u on Malaita and Vella Lavella in Western Province. In Vanuatu households were surveyed in Port Vila (the capital), Mangalilu/Lelepa Island on the main island of Efate, Luganville (the second largest town) and Hog Harbour in Espiritu Santo, Baravet on Pentecost and Mota Lava in the Banks (see Chapter 1 for further details).

Internal migration and the manner in which Solomon Islanders and Ni-Vanuatu shift between rural and urban locations as well as their customary ties to land and access to the sea have clearly affected the experience of shocks in these countries. An important feature of our findings, which perhaps reflect the fact that the majority of households surveyed are located in rural areas, shows that there is a significant migration to rural areas (from urban areas as well as from other rural areas). It is important to stress, however, that this does not contradict the fact that rates of urbanization are increasing in these countries. It does, though, imply a high level of 'circular migration' – short term seasonal migration to and across rural areas possibly for employment and family reasons. Moreover, return migration could result from the fact that urban households are found to be relatively more vulnerable to economic shocks than their rural counterparts. Finally, we also find that remittances, particularly in the form of goods (such as clothes and food), are a very important tool used to deal with vulnerability in the Solomon Islands and Vanuatu. Such flows have also been shown to be important in other developing economies, particularly in Latin America (Jennings and Clarke, 2005; Acosta et al., 2006, 2007).

The remainder of this chapter is structured as follows. Section 4.2 presents a discussion on the theory and practice behind the choice to migrate. Section 4.3 presents evidence on migration patterns in Melanesia and the Solomon Islands and Vanuatu in particular. Section 4.4 uses data from the research fieldwork to examine whether urbanization has lowered vulnerability while Section 4.5 examines migration and the importance of household remittances. Finally, Section 4.6 concludes.

4.2 The Motivations for Migration

There has been great interest in understanding migration and labour market flows since the onset of the industrial revolution. There are, in particular, two theoretical models that are often employed to analyse urbanization and internal migration in developing economies. The first is the Lewis model or Dual Sector model (Lewis, 1954) and the second is the Harris–Todaro model (Harris and Todaro, 1970). Both theories model a transition by which labour from a traditional agricultural sector migrates to a modern industrial sector due to wage differentials. Over time, the transition of workers across sectors will drive down real wages in the modern sector relative to the traditional sector. Eventually, the wage rates of the two sectors equalize, increasing productivity and wages in agriculture whilst driving them down in manufacturing. Note however, that while these two models are useful in explaining internal migration internationally, they are less relevant in explaining migration in Melanesia. In Melanesia, migration is shaped more by traditional obligations to extended families as well as by ties to land and access to the sea upon which traditional economic activity is based.

International evidence has highlighted that migration is most often a reaction to 'push' or 'pull' factors. Push factors are, for instance, environmental factors, a lack of employment opportunities and cost-of-living problems in the region of origin. In particular, environmental push factors such as cyclones and floods deserve a special mention in Melanesia, given the region's exposure to natural disasters. Environmental problems often drive people to migrate to geographically higher and secure places and to urban areas from rural locations and outer islands. Pull factors are often driven by the desire to live closer to and support one's family, to access better employment opportunities or to more easily benefit from essential services. However, migrants in the Pacific often return to rural locations due to economic hardships in destination regions (Brown and Jimenez, 2008; Clarke, 2009).

Migration often leads to urbanization, which allows for the urban sector to become an additional resource for *rural* development (Skeldon, 1997). For example, remittances (money and goods) from family members living and working in other countries overseas are very important for some Pacific island economies (Connell and Brown, 2005). In particular, Polynesian countries have benefitted from international migration due to the availability of policies that facilitate their access to other countries' labour markets. People from Melanesia have not benefitted from such access to other countries' labour markets and migration has therefore been dominated by internal movements (Maclellan, 2008; Hammond and Connell, 2009; Gibson and McKenzie, 2011). Migration – domestic or international – does facilitate the transfer of resources from one region (usually the rural) to another (usually the urban). However, migration can also result in perverse outcomes by depriving villages of their most educated and energetic members. For example, seasonal worker schemes in the Pacific have been found to deprive those left at home, *vis-à-vis* other households in the community, owing to the fact that the most productive member of the household is absent. Additionally, rapid urbanization, resulting from rural-urban migration in the Pacific has resulted in serious environmental, economic and social problems (Connell and Lea, 2002).

The implications of the traditional economic models of labour mobility, as well as previous empirical evidence from the Pacific, are important. However, previous studies have failed to systematically analyse the migration of household members in the aftermath of economic and other shocks. Focusing on the Solomon Islands and Vanuatu, in particular, is also of great interest because of increasing rates of urbanization as well as the stronger family and customary ties to the land and sea upon which economic activity has been traditionally based.

Population movement in the Solomon Islands and Vanuatu is fluid. Consistent with previous studies, such as Bedford (1973), this chapter finds that migration is often 'circular' with people migrating from rural to urban centres alongside migration occurring from urban to rural areas. In addition, there is often movement across rural locations. This indicates that while urbanization is certainly a feature of modern Melanesia, rural and urban populations are dynamic. Within Melanesia, a better term therefore to describe this movement might be *mobility* rather than migration. Mobility more accurately describes shifting population movements because of Melanesian *kastom*.

Kastom are the traditions that govern obligations, behaviours and relationships between families and communities within Melanesia. *Kastom* also involves physical manifestations of these relationships and obligations in dance, dress, food, song and history. An important aspect of *kastom* that is highly relevant to mobility within Vanuatu and Solomon Islands are the ties to customary land ownership. While specific rights differ between communities, a common characteristic is the importance of land (and sea) and rights to access over this land. This customary ownership of land motivates very strong personal and community ties to land as well as a sense of personal identity. Leach et al. (2012) find that amongst tertiary students within the Solomon Islands and Vanuatu, whilst national affiliations were strong, young educated people felt much closer to their village than their island or region/province. This emotional attachment to land and village is an important reason why mobility better explains human movements within Vanuatu and Solomon Islands. Such customary obligations and ties do not weaken when people shift from their home villages to urban centres.

Other obligations under *kastom* include the care and support of extended family members (*wantok*). Within rural communities, semi-subsistence agriculture is the mainstay of local economies. The production of food and housing is reliant on the access to and use of land. Provision of assistance to one's own immediate family and larger *wantok* is possible through the 'garden economy' and the ability of households to grow a lot of the food that they consume. This study found that for the majority of respondents who had shifted from rural locations to urban centres, dependence on the garden economy and support from their *wantok* remained or increased and was an important aspect of their resilience. There is limited access to land for gardens in some urban areas. Moreover, employment opportunities are still limited in many urban parts of the Pacific and some migrants end up relying on food and other items sent to them from rural areas (Connell and Brown, 2005).

4.3 Migration and Mobility in Melanesia

As in most developing countries, households in Melanesia have experienced sustained levels of internal mobility (Lindstrom, 2012; Bonnemaison, 1984). This is clearly confirmed by the household survey. Defining a migrant as anyone who moved internationally, within an island or to a different island, 45 per cent of respondents in the Solomon Islands and 41 per cent in Vanuatu can be said to have migrated in their lifetime. Almost half of respondents who migrated in each country migrated from a different island and the other half from another part of their current island of residence. Significantly, only 2 per cent migrated to the Solomon Islands or Vanuatu from a third country highlighting that Melanesian mobility is largely a domestic phenomenon.

Table 4.1 Push and pull mobility factors in the Solomon Islands and Vanuatu

		Solomon Islands		Vanuatu	
		Households	(%)	Households	(%)
Push Factors					
	Environment	16	5	11	3
	Employment	16	5	14	4
	Affordability	8	2	13	4
	Family	8	2	23	7
Total (Push)		48	17	61	19
Pull Factors					
	Employment	48	15	25	8
	Affordability	36	11	53	16
	Family	135	41	123	37
Total (Pull)		219	67	201	61

Source: The authors.

Table 4.1 reviews some of the push and pull factors identified by household respondents. The table decomposes push and pull factors into the following categories: (i) environment; (ii) employment; (iii) affordability and; (iv) family. Environmental factors refer to natural disasters, employment refers to the search for labour market opportunities (including educational opportunities), affordability refers to the need to reduce the amount of money spent or to improve living conditions, and family refers to reasons such as marriage, proximity to family or disputes. Perhaps because a higher proportion of people in rural than urban areas were surveyed, the data indicates that the majority of respondents identified 'pull', rather than 'push' reasons for their mobility.

Table 4.1 highlights that mobility in both countries is driven by return migration for family reasons. This corroborates evidence in Gibson and McKenzie (2011), who find that return migration is strongly linked to family and life-style reasons for New Zealand, Papua New Guinean and Tongan migrants. The remaining reasons for mobility include traditional economic reasons, such as employment and seem to be evenly distributed amongst respondents.

Data from the World Bank (2012) suggests that over the past decade, the proportion of people living in urban areas has increased by 29 per cent in the Solomon Islands and 25 per cent in Vanuatu. This gives impetus to the argument that the Solomon Islands and Vanuatu, as well as other countries in the region, have experienced very high rates of urbanization (Connell, 2011). Nevertheless, our survey, which documents mobility patterns at the household level, finds some evidence to suggest that some migrants are leaving urban areas to resettle in rural ones.

Table 4.2 Push and pull mobility factors in rural and urban areas

		Rural		Urban	
		Households	(%)	Households	(%)
Push Factors					
	Natural disasters	23	5	4	2
	Employment	20	4	10	5
	Affordability	7	2	14	7
	Family	14	3	17	9
Total (Push)		64	14	45	24
Pull Factors					
	Employment	52	11	21	11
	Affordability	62	13	27	14
	Family	207	44	51	27
Total (Pull)		321	69	99	52

Source: The authors.

Table 4.2 provides figures on push and pull mobility factors for urban and rural areas. As in Table 4.1, there seems to be a fairly even dispersion between the push factors in both rural and urban areas. However, migration of people currently living in rural areas seems to be driven by the wish to return home. Therefore, although mobility patterns generally follow flows predicted by classical models, we do find some evidence to suggest that there are significant flows of return-migrants to rural areas. This, in turn, hints that earnings inequality between urban and rural sectors is still small, although probably widening. Data from the household survey indicates that the urban/rural earnings ratio is 1.22 and 1.11 for the Solomon Islands and Vanuatu, respectively.

Urbanization is a feature of modern economic development, and whilst this applies to the Solomon Islands and Vanuatu, it is also important to note that the 'stock' of the urban populations (whilst growing) is also constantly 'refreshing' itself with a through-put of migrants moving between rural and urban locations. Amongst other reasons, this mobility is partially explained by Solomon Islanders and Ni-Vanuatu responding to household shocks and by re-engaging with the customary garden economy. These issues are explored further in the next section.

4.4 Shocks, Migration and Urbanization

The discussion above suggests that family is dominant in driving internal mobility in the Solomon Islands and Vanuatu. Notwithstanding, it is interesting to examine whether households experiencing economic shocks as well as increments in the prices of food and fuel migrated in recent years. Household data reveal that 41 and 44 per cent of rural and urban households, respectively, experienced an income shock (defined as an unexpected fall in household income over the preceding two years). Therefore, we cannot conclude that there is a significant difference between urban and rural households in this regard.[1]

However, households in urban areas have experienced significantly more price shocks. For instance, while 10 per cent of rural households found that buying food has become much harder, 18 per cent of urban households encountered the same problem. Similarly, 7 per cent of rural households found that buying fuel has become more difficult, compared to 12 per cent of urban households. The latter suggests that urbanization has made urban households relatively more vulnerable to price shocks than their rural counterparts.

Mobility can be used as a tool by households in the Solomon Islands and Vanuatu to hedge against shocks. Given the high rates of urbanization present in Melanesia, it is relevant to examine further whether households in urban areas are more vulnerable to shocks and their impacts. Consideration of this begins

1 There is no standard definition of an economic shock. It could be classified as a household either experiencing reduced employment (job loss or reduced hours), reduced demand for goods sold, reduced supply of goods sold, reduced remittances or an increase in household size. If this definition is used, urban households in the Solomon Islands and Vanuatu are more likely to experience an economic shock than rural households.

with an analysis of food security. The household survey asked respondents to identify situations where food security might have been an issue. Table 4.3 identifies that urban households are far more vulnerable to food insecurity than rural ones, suggesting that urbanization may have perverse effects toward long-term poverty reduction. This must be evaluated with the acknowledgement that households in squatter settlements in urban areas were targeted by the survey, in which households sometime had very little access to land and a garden.

Table 4.3 Food security in urban and rural regions (per cent of households)

Household Food Scarcity (Whether there was a time in the past two years when the household was not able to afford food)			
	Don't know	True	False
Rural	0	48	52
Urban	2	73	26
Total	1	56	43
Child Food Scarcity			
Rural	0	19	81
Urban	1	40	59
Total	0	27	73
Cheaper Substitutes			
Rural	0	57	43
Urban	1	76	23
Total	1	64	34

Note: Scarcity refers to either the household or children within the household not being able to afford food. Cheaper substitutes refer to the household switching their diet toward cheaper food substitutes.

Source: The authors.

Rates of food insecurity in the Solomon Islands and Vanuatu in which a garden economy dominates are alarming. Table 4.3 shows that 73 per cent of urban households faced situations where they were not able to afford food (scarcity), compared to 48 per cent of rural households. Additionally, 40 per cent of urban households encountered situations where they feared not being able to feed their children, as opposed to 19 per cent of rural households. Finally, Table 4.3 shows that 76 per cent of urban households switched their consumption bundles toward cheaper (possibly less nutritional) food stuffs, compared to 57 per cent of rural households. These findings highlight the importance of the role of the garden economy in Melanesia as a tool to hedge against the shocks associated with poverty. Additionally, it highlights the prominent role that food remittances might have in these economies. These issues are discussed in more detail below.

The household survey also asked whether adults or children experienced hunger, defined as going a day without food, over the last 12 months. Our data shows that 23 per cent of adults experienced hunger in urban areas, relative to 11 per cent in rural areas. Similarly, 11 per cent of children in urban areas experienced hunger (going without food for a day), compared to only 5 per cent in rural areas. Table 4.4 summarizes data on the frequency of these experiences for both adults and children in order to ascertain the severity of the problem in each region.[2] The table shows that 30 per cent of adults that experienced hunger in urban areas did so every month, compared to 21 per cent of in rural areas. Moreover, adults in rural areas are found to be more likely than adults in urban areas to experience hunger in only one or two months out of the year. Similar patterns emerge when addressing this issue in regards to children. Of the children in urban areas that experienced hunger, 53 per cent did so every month, relative to 24 per cent in rural areas. Overall, this suggests that families in rural regions are better placed to hedge against food insecurity. The main reason behind this is that these families are better placed to use gardens to grow their own food in times of scarcity – the household data indicate that 94 and 80 per cent of households in urban and rural areas, respectively, use gardens to grow their own food.

Table 4.4 Percentage of households reporting that adults and children have gone a day without food during the last 12 months

	Adults		
	Every month	Some months	1 or 2 months
Rural	21	41	38
Urban	30	54	13
	Children		
Rural	24	59	17
Urban	53	41	18

Source: The authors.

2 Note that only a small number of households responded to the questions summarized in Table 4.4, the sample covers 132 adults and 61 children.

Food security is just one of the many ways we can make inferences as to the levels of vulnerability evident in Melanesian households in the presence of higher urbanization. Table 4.5 presents data relating to household access to food, fuel, money, education, healthcare, water and sanitation, roads and security. Respondents were asked whether access to these goods and services have improved, improved significantly, stayed unchanged, worsened or worsened significantly over the past two years. Table 4.5 summarizes the findings in terms of whether access improved (became easier) or worsened (became harder).

Table 4.5 Percentage of households reporting changes in access to goods and services

		Rural	Urban
Food	Easier	44	28
	Harder	26	42
Fuel	Easier	16	11
	Harder	64	69
Money	Easier	28	19
	Harder	54	60
Education	Easier	50	44
	Harder	32	36
Health	Easier	48	36
	Harder	30	39
Water/Sanitation	Easier	29	25
	Harder	37	51
Roads	Easier	28	17
	Harder	45	48
Security	Easier	27	21
	Harder	26	43

Source: The authors.

Overall, the respondents noted that obtaining access to these goods and services has become more difficult in urban than rural areas. For example, 42 per cent of respondents indicated that accessing food has become more difficult in urban areas compared to 26 per cent of respondents in rural areas. Similarly, 69 per cent of respondents noted that accessing fuel is become harder in urban areas, while 64 per cent indicated this to be the case in rural areas. Similar patterns emerge in accessing money, security, roads, and water and sanitation. Interestingly, educational attainment has improved in both regions, although evidence here suggests that improvements in rural areas have been greater than those in urban regions. Finally, access to healthcare has improved in rural areas and worsened in urban areas – 48 per cent of rural households indicated that access to healthcare has improved, while 39 per cent of urban households suggested it hard worsened (while 25 per cent of these households noted that access to healthcare remained unchanged in urban areas over this period).

4.5 Mobility and Remittances

One of the most important direct consequences of mobility and urbanization is the flow of remittances that follow. Remittances directly increase the income of recipients, which helps smooth household consumption, especially in response to adverse events, such as a natural disaster, crop failure or a health crisis. Moreover, by raising income, remittances also appear to be associated with increased household investments in education, entrepreneurship and health – all of which engender a high social return in most circumstances (World Bank, 2006). Skeldon (1997) argues that remittances are more likely to be associated with international (rather than internal) migration, however cash and goods flows have been found to be an intrinsic component of internal mobility in Melanesian countries.

The household survey data indicates that the remittances households received in the form of money make up a very small proportion of total household income. This is true in both urban and rural areas. It is found that remittances to rural households account for just 2 per cent of total household income in both the Solomon Islands and Vanuatu. Similarly, remittances to urban households accounted for just 1 per cent of their total household income. These figures are small compared to other surveys, which have noted much larger percentages, particularly in the wider Pacific (World Bank, 2006). It is also important to ascertain whether these flows have changed in the last few years. The survey indicates that of the 60 households which responded to the question of whether remittances increased/decreased/stayed the same, 24 reported that they stayed the same, 19 said they received more now and 17 said they receive fewer remittances. Overall, we cannot conclude that inflows of cash remittances are significantly allowing Melanesian households in the sample to adequately deal with economic shocks. While it is common for households to receive remittances, findings from the household survey suggest that the amounts received in the form of money are very small relative to the total income of the household.

An important feature of migration in developing countries is that often households will both send and receive remittances through somewhat complex social networks. The household survey finds that values of remittance outflows are significantly larger than their inflow counterparts. For rural households, outflows of remittances are found to account for 11 per cent and 24 per cent of household income in the Solomon Islands and Vanuatu respectively. For urban households, outflows of remittances are found to account for 12 per cent and 18 per cent of household income respectively.

These findings can be explained by three alternative hypotheses. First, households in these regions are net remittance givers. Second, households are quite possibly overstating the amount of money they give relative to the amount received. Third, households that send remittances may be adjusting their estimated expenditures to include outflows of not only money, but also consumer goods.

Remittances in developing countries can often take the form of goods in addition to cash (Posso, 2012). As such, the actual monetary value of remittances can often understate the real flow. In order to account for this, we redefine remittances as outflows/inflows of not only cash, but also clothing and food items. While it is impossible to determine the actual monetary value of food and clothing transfers, it is possible understand whether these transfers are made. Table 4.6 presents the proportion of households that send and receive this broader definition of remittances in the Solomon Islands and Vanuatu. The data are presented by whether households are located in urban or rural areas. Additionally, the table presents information on whether respondents found that remittances outflows or inflows increased, decreased or stayed the same during the last two years.

Overall, Table 4.6 shows that remittance patterns do not differ significantly between rural and urban areas. Additionally, we find that food remittances are particularly more important in these two countries than money and clothing remittances. For instance, 86 per cent of urban households in Honiara send food remittances, compared to 74 per cent that send money and 59 per cent that send clothes. Similar patterns are evident in Vanuatu, although the proportions of households that send remittances are significantly higher. Remittance inflows of food are significantly larger than clothes and money in both countries and in both urban and rural areas. For example, 81 per cent of rural households in the Solomon Islands received remittances in the form of food, compared to only 61 and 42 per cent that received remittances in the form of money and clothing, respectively. As above, similar patterns are evident in Vanuatu. Overall, flows of food between households may be significantly allowing the vulnerable to become more resilient to economic shocks. This highlights the importance of traditional custom in these small island communities.

Recent hikes in the prices of food and fuel as well as the GEC may have significantly affected remittance flows internationally. In order to establish whether this is the case in the Solomon Islands and Vanuatu, households were asked if remittance outflows/inflows increased, decreased or stayed the same over the last two years. These data are summarized in the last three columns of Table 4.6. Overall, the data suggest that remittances are on the decline in the two countries concerned. The proportion of households reporting a decrease in the amount of money, food and clothes they remit as well as the amount they receive exceeds the proportion of households reporting an increase. Reported falls are particularly high in Vanuatu and potentially highlight a deterioration of the custom or traditional economy in which households look after their extended family during times of need. It could also indicate that increasing monetization and rising prices are leading to financial stress in an increasing number of households and they have simply become unable to assist others in need.

Table 4.6 Percentage of households sending/receiving remittances and whether amounts have increased/decreased

Solomon Islands	Region		Change (Last 2 Years)		
	Urban	Rural	Increase	Same	Decrease
Outward remittances					
Money	74	74	24	40	36
Food	86	89	23	49	28
Clothes	59	56	24	47	29
Inward remittances					
Money	55	61	11	47	42
Food	70	81	12	56	32
Clothes	25	42	15	53	32
Vanuatu					
Outward remittances					
Money	89	82	30	26	44
Food	96	92	30	33	37
Clothes	87	83	27	30	43
Inward remittances					
Money	67	71	22	30	48
Food	85	86	20	36	44
Clothes	66	69	18	35	47

Source: The authors.

Remittances in the Solomon Islands and Vanuatu do differ from other regions, with a heavy emphasis on goods relative to financial remittances. Again, this relates directly to the social obligations associated with *wantoks* and the strong connection to the *kastom* of the garden economy as the basis for meeting social obligations but also that this subsistence economic behaviour continues to feature as a common aspect of life and an important source of resilience.

4.6 Conclusion

Mobility in Melanesia, particularly Solomon Islands and Vanuatu, is a better descriptor than migration to understand the patterns of human movement within this part of the world. While urbanization in these two countries continues to grow, there remains a circular flow between rural and urban locations as Solomon Islanders and Ni-Vanuatu move to and from the rural and urban sectors. This mobility is a reflection of *kastom* that involves strong ties and access to land that continues to underpin the traditional garden economy that remains the mainstream economic activity of many in these two countries. Certainly, *kastom* does not lessen when people relocate to urban centres. Moreover, the garden economy and remittances of food have served as a buttress against recent economic shocks and have been a fundamental aspect of the resilience of those in urban centres. This resilience and the reliance of the garden economy has resulted in people shifting 'back' to rural locations so they can directly participate in the garden economy through accessing their customary access to land and the sea or by relying on *wantok* obligations and having (extended) family members remit food from rural locations to their urban homes.

The value and importance of *kastom*, *wantok* and the garden economy should be given prominence by policy-makers and those involved with planning and initiating community development interventions when planning for and responding to shocks. Whilst urbanization continues to characterize Melanesia, the mobility between urban centres and rural locations that is predicated upon strong customs, results in economic resilience remaining strongly tied to traditional economic activities and social obligations to extended family members.

References

Acosta, P., Calderon, C., Fajnzylber, P. and Lopez, H. (2006), Remittances and Development in Latin America, *The World Economy,* 29(7): 957–87.

Acosta, P., Calderon, C., Fajnzylber, P. and Lopez, H. (2007), What is the Impact of International Remittances on Poverty and Inequality in Latin America?, *World Development*, 36(1): 89–114.

Bedford, R. (1973), *New Hebrdies Mobility: A Study of Circular Migration* (ANU Department of Human Geography, Australian National University: Canberra).

Bonemaion, J. (1984), The Tree and the Canoe: Roots and Mobility in Vanuatu Societies, *Pacific Viewpoint*, 27: 7–15.

Brown, R.P.C and Jimenez, E. (2008), Estimating the Net Effects of Migration and Remittances on Poverty and Inequality: Comparison of Fiji and Tonga, *Journal of International Development*, 20: 547–71.

Clarke, M. (2009), Economic Growth and Outlook, *Background Paper for Pacific Economic Survey* (Australian Agency for International Development: Canberra).

Connell, J. (2011), Elephants in the Pacific?: Pacific Urbanisation and its Discontents, *Asia Pacific Viewpoint,* 52(2): 121–35.

Connell, J. and Brown R.P.C. (2005), *Remittances in the Pacific: An Overview* (Asian Development Bank: Manila).

Connell, J. and Lea, J. (2002), *Urbanisation in the Island Pacific: Towards Sustainable Development* (Routledge: London and New York).

Gibson, J. and McKenzie, D. (2011), The Microeconomic Determinants of Emigration and Return Migration of the Best and Brightest: Evidence from the Pacific, *Journal of Development Economics,* 95(1): 18–29.

Hammond, J. and Connell, J. (2009), The New Blackbirds?: Vanuatu Guestworkers in New Zealand, *New Zealand Geographer*, 65: 201–10.

Harris, J.R. and Todaro, M.P. (1970), Migration, Unemployment and Development: A Two-Sector Analysis, *American Economic Review,* 60(1): 126–42.

Jennings, A. and Clarke, M. (2005), The Development Impact of Remittances to Nicaragua, *Development in Practice,* 15(5): 685–91.

Leach, M., Scambary, J., Clarke, M., Feeny, S. and Wallace, H. (2012), Attitudes to National Identity among Tertiary Students in Melanesia and Timor-Leste: A Comparative Analysis, SSGM Discussion Paper 2012/8 (State Society and Governance in Melanesia, Australian National University: Canberra).

Lee, H. (2009), Pacific Migration and Transnationalism: Historical Perspectives, in H. Lee and S.T. Francis (eds), *Migration and Transnationalism: Pacific Perspectives* (Australian National University Press: Canberra).

Lewis, A.W. (1954), Economic Development with Unlimited Supplies of Labor, *Manchester School of Economic and Social Studies,* 22: 139–91.

Lindstrom, L. (2012), Urbane Tannese: Local Perspectives on Settlement Life in Port Vila, *Journal de La Societe des Oceanistes*, 2(133): 255–66.

Maclellan, N. (2008), Pick of the Crop. New Zealand Opens up to Seasonal Workers, *Pacific*, 33(2): 14–21.

Posso, A. (2012), Remittances and Aggregate Labour Supply: Evidence from 66 developing nations, *The Developing Economies,* 50(1): 25–39.

Skeldon, R. (1997), *Migration and Development: A Global Interpretation* (Longman: London).

World Bank (2006), *Global Economic Prospects 2006: Economic Implications of Remittances and Migration* (World Bank: Washington).

World Bank (2012), *World Development Indicators Online Database* (World Bank: Washington).

Chapter 5
Vulnerability to What? Multidimensional Poverty in Melanesia

Matthew Clarke, Simon Feeny and Lachlan McDonald

5.1 Introduction

Defining poverty as a lack of income is intuitively attractive and dates back to the earliest work on poverty in England during the nineteenth century (see Booth, 1887; Rowntree, 1902). From an individual's experience, income affords us the freedom to purchase our basic needs and many of our desired wants. Further, having money facilitates choices about the things we desire (whether they are good or bad for us). In economic terms, our utility (or happiness) increases as consumption increases by the simple fact that purchasing a particular commodity reveals our preference (belief) that this commodity will increase our utility. Thus the more we purchase, the greater our utility. As having unlimited desires is said to be a human characteristic, an increase in income therefore increases our ability to maximize our utility. Conversely, having less money reduces our ability to consume and lowers our utility. At the extreme, an income below a certain level means that even the basic needs of food, shelter and clothing cannot be adequately met. Then an individual, or household, can be said to be experiencing poverty.

However, the relevance of income-based poverty to Melanesian countries is disputed. With strong family and social support networks, outright destitution has been rare and the term hardship is preferred over poverty (Abbott and Pollard, 2004). With ongoing debates about the specific nature of poverty in these contexts, assessing poverty and developing policy responses is challenging. It has been considered, for example, that Melanesians live in an environment of 'subsistence affluence', unburdened by modern economic problems. As such, income-based assessments of poverty have largely been considered irrelevant or, at best, unsuitable. However, with the considerable changes wrought by increasing monetization and urbanization 'subsistence affluence' has arguably become redundant for many Melanesian societies. Malnutrition and hunger exist, with households relying on cheap, sometimes poor quality imported food and sometimes missing meals. Monetization is increasing the importance of an income to enable the payment of school fess and the purchase of staple household goods and services in order to meet the basic needs of the family.

Poverty is now universally recognized as being multidimensional in nature. A multidimensional approach is certainly relevant to Melanesia where it would be inadequate to measure poverty without consideration of access to health, education, clean water and sanitation, housing and access to markets, as well as other factors which impact on a household's ability to meet its cultural obligations. Vulnerability to poverty – the likelihood, or risk, of being poor or falling into poverty in the future – is an increasingly important concern for policymakers in Melanesia.

This chapter uses the data from a household survey conducted in the Solomon Islands and Vanuatu in 2010–11 to calculate a Multidimensional Poverty Index (MPI). The survey was conducted in 12 diverse communities in the Solomon Islands and Vanuatu and was specifically designed to measure the incidence and depth of poverty (see Chapter 1 for further details on the household survey). We replicate the MPI, developed by the Oxford Poverty and Human Development Initiative (OPHI) (Alkire and Foster, 2011a). The MPI measures a number of deprivations that a household experiences. More specifically it calculates the percentage of households that experience overlapping deprivations in three dimensions: education, health and living conditions. The index is widely regarded as a useful measure of poverty and country level values are now published in the annual Human Development Reports of the United Nations Development Program (UNDP). However, as the creators of the index acknowledge, it can be modified to better reflect the living conditions and livelihoods of specific country contexts. The chapter therefore proceeds by augmenting the MPI with information on access to a produce garden, health, education services and local markets, important to the context of Melanesian communities. The augmented index is referred to as the Melanesian MPI (or MMPI). In the case of Vanuatu, the analysis complements MNCC (2012) which examines alternative indicators of wellbeing. The MNCC report focuses on self-reported happiness and life satisfaction and correlates these scores with resources access, measures of culture and community vitality.

The remainder of this chapter is structured as follows. Section 5.2 outlines how poverty is measured before Section 5.3 examines the findings from constructing the MPI and the tailored MMPI using the household survey data. Finally, Section 5.4 concludes with some policy recommendations.

5.2 Measuring Poverty and Well-Being

The shift from understanding poverty as being based largely on income deficits to be more encompassing of different dimensions of well-being has now been universally accepted. Sen has been influential in shifting this conceptual understanding, along with work such as Nussbaum's central human capabilities, Doyal and Gough's intermediate human needs, and Narayan, et al. identifying axiological needs, among many others (Sen, 1984, 1993; Nussbaum, 1988, 1992, 2000; Doyal and Gough, 1991; Narayan, et al., 2000). In response to this conceptual shift, there have been three significant developments in assessing poverty and human well-being: the Human Development Index (HDI), the international community's commitment to the Millennium Development Goals (MDGs) and the recently devised MPI. Each is discussed in turn.

5.2.1 Human Development Index

The HDI is a composite index based upon Sen's concept of capability (UNDP, 1990; 2011). Combining proxy indicators associated with education, health and living standards, the HDI was initially established to counter the hegemonic status of national income as the default measure of human well-being. As demonstrated by Table 5.1, there are significant differences in the HDI across the Pacific region. At one end, Palau and Tonga are classified as having high human development (scores in 2011 of 0.782 and 0.704 respectively), putting them on par with the Latin American countries of Uruguay and Mexico and European Montenegro and Romania. At the lower end of human development in the Pacific, are Papua New Guinea and the Solomon Islands (with scores in 2011 of 0.466 and 0.510 respectively). Indeed, Papua New Guinea is ranked 153 (out of 186) countries and is one of four non-African countries at the bottom-end of the HDI rankings (along with Nepal, Afghanistan and Haiti). This suggests that Papua New Guinea significantly lags behind its regional neighbours Kiribati, Samoa, Fiji, and Vanuatu which are all classified as experiencing medium human development.

Generally countries across the Pacific, including Papua New Guinea, have seen improvements (albeit sometimes small) in their HDI scores over time. This indicates at least some progress is being made in improving well-being.

Table 5.1 Human Development Index for selected Pacific island countries

	2000	2005	2010	2011
Federated States of Micronesia		0.633	0.635	0.636
Fiji	0.668	0.678	0.687	0.688
Kiribati			0.621	0.624
Palau	0.774	0.788	0.779	0.782
Papua New Guinea	0.423	0.435	0.462	0.466
Samoa	0.657	0.676	0.686	0.688
Solomon Islands	0.479	0.502	0.507	0.510
Tonga	0.681	0.696	0.703	0.704
Vanuatu			0.615	0.617
Pacific	0.479	0.490	0.508	0.511

Source: UNDP (2011).

5.2.2 Millennium Development Goals

In 2000, 189 nations committed themselves to the achievement of a number of development targets known as the MDGs. In so doing the international community indicated an intention to address the poverty affecting billions of the world's population. Emanating from a number of international conferences held throughout the 1990s, the MDGs are designed to address many of the multidimensional aspects of poverty. The eight goals: (i) eradicating extreme income poverty and hunger; (ii) achieving universal primary education; (iii) promoting gender equality; (iv) reducing child mortality; (v) improving maternal health; (vi) combating HIV/AIDS, malaria and other diseases; (vii) ensuring environmental sustainability; and (viii) developing a global partnership for development are to be assessed against 18 targets and 48 indicators. The international community set 2015 as the year by which these global targets are to be achieved using 1990 as a baseline.

The value of the MDGs is not only that they identify a series of goals with agreed targets and indictors, but also that they set a timeline for their achievement. Indeed, this was the first time the international community had set itself a date by which improvements in well-being would be achieved and for which members would be held to account.

The importance of the MDGs cannot be overstated. There is little time before 2015 – the end of the MDG timeline – and it is increasingly clear which goals will be achieved and which will be missed. The Asia-Pacific region, for instance, is often reported to be making good progress towards MDG achievement but the region is extremely diverse and analysis at the regional level masks significant differences in the progress of individual countries.

It has become common practice to assess country-level progress against the global MDGs. This often results in a summary statement of whether a country has 'achieved' a goal, is 'on-track' or 'off-track' in terms of its progress, providing a quick overview of how a country is progressing towards the MDGs and targets. Such an approach, using national averages, masks gendered progress, sub-regional or ethnic disparities and other inequities and has stimulated criticism of the assessment of specific countries against global targets (Feeny and Clarke, 2009; Vandermoortele, 2009, 2011).

Using this approach, however, it is increasingly apparent that Pacific countries, in general, have made very limited progress towards achieving the development targets. According to the Pacific Island Forum Secretariat (PIFS), the 15 Pacific Island Forum member nations are off-track to achieve all goals (PIFS, 2011). Note, however, that progress on the goal in regard to global partnerships (MDG 8) is not explicitly assessed as its premise is that 'the developing nations would focus on achieving the first seven goals, while developed countries would support these efforts through increased aid, fairer market access and debt relief, as well as ensuring affordable essential drugs and information communications and technology' (PIFS, 2011, p. 8).

The Polynesian sub-region is the best-performing of the Pacific, on-track to achieve the global targets of MDG 2 (education), MDG 4 (child mortality reduction), MDG 5 (maternal health) and MDG 7 (environmental sustainability). For Micronesia as a whole, the results are mixed for all goals, other than eradicating extreme income poverty and hunger (MDG 1) for which the entire regional group is off-track. Melanesia is performing worse by these measures, and as a whole is off-track to achieve any of the global targets. However, if Papua New Guinea is excluded, then the prognosis for the rest of Melanesia is less bleak, being off-track to achieve gender quality, on-track to achieve child mortality reduction and experiencing mixed results for the remaining goals.

Off-track for them all, Papua New Guinea is the least likely of all countries within the region to meet any of the goals. In this regard, the Solomon Islands is the second poorest performing country in the region. The overall analysis remains very similar, in terms of countries at risk and countries doing well when using the slightly different Australian Agency for International Development system to categorize progress (AusAID, 2009).

However, as noted, the discussion of individual nation goal achievement is not appropriate. The targets were, in the majority, based upon the extrapolation of global trends rather than on progress at an individual country level (Vandermoortele, 2009). The relevance of the MDGs to the Pacific region also requires consideration. It may be that the goals are not specifically suited to the economic, social and geographical characteristics of these small-island states. Indeed, the original concept of the goals as global targets was lost immediately following their adoption so that much attention is now paid to distinct countries, rather than these countries being assessed against their contribution towards the achievement of global targets (Vandermoortele, 2011).

Thus, while it might be sound for the world as a whole to aim to halve the proportion of the population living on less than US$1.25 per day, extending previous trends at the global level, is it an appropriate target for a country without a history of such improvements? Similarly, unique cultural circumstances may make the focus on certain goals and targets inappropriate for some regions or places. It is also important to note that it is possible for countries to tailor the MDGs to their own circumstances. As development goals are important, rather than reject them because they are not completely aligned to existing development plans, or are unlikely to be achieved, the answer lies in tailoring them to specific country contexts. What essentially matters is the existence of appropriate, mutually agreed targets that governments and the international community can work towards (Feeny and Clarke, 2009).

A tailoring of the MDGs has been supported by Pacific nations within the Port Vila Declaration on Accelerating Progress on the Achievement of the MDGs. Papua New Guinea, for example, has tailored the goals, making some targets less ambitious but more realistic for achievement by 2015 (Feeny and Clarke, 2009). The purpose of this discussion is to point to the necessary nationalization of international development measures and plans.

Data availability is a considerable constraint is assessing well-being within the Pacific. With scarce resources available to collect national data from small, geographically dispersed populations, the ability of small-island states to collate and analyse data is limited. For many Pacific nations, data collection has been given a low priority due to both its expense and high technical requirements (Feeny and Clarke, 2008, 2009; PIFS, 2011). Information is often out dated, incomplete or entirely unavailable. This data difficulty is long-standing and recognized by donors and Pacific Island governments alike. Real progress towards the MDGs by Pacific Island nations is therefore difficult to measure as doing so requires accurate and comparable data both across time (inter-temporal) and across space (interspatial). The only two Pacific countries with data available for the MDG headline indicator of halving the proportion of population on less than US$1.25 per day are for the Federated States of Micronesia (FSM) and Papua New Guinea.

5.2.3 The Multidimensional Poverty Index

Approaches to measuring both development and poverty have widened to incorporate the knowledge that low income does not account for the array of possible achievements that characterize development, nor for the array of deprivations that characterize poverty. Following Sen's conceptual work and utilizing improved availability of data, there has been increased interest within the literature on moving beyond the common monetary-based headcount of poverty to develop a multidimensional measure (Kakwani and Silber, 2008). Yet, while the multidimensionality of poverty is no longer disputed, there is not a full consensus on how multiple dimensions should be captured and assessed. Nor will there ever be.

The MPI is perhaps the best known of recent efforts in this field, and is certainly the most widely applied, having now been estimated for more than 100 countries (Alkire and Foster, 2011a). Alkire and Foster took as their starting point the contention of Bourguignon and Chakravarty that a 'multidimensional approach to poverty defines poverty as a shortfall from a threshold on each dimension of an individual's well-being' (Bourguignon and Chakravarty, 2003, p. 25). Thus, the MPI considers three equally weighted dimensions of poverty through ten indicators. However, Alkire and Foster qualify this since, 'the poverty status of a person is unaffected by certain other changes in achievements. For example, a poor person can never rise out of poverty by increasing the level of non-deprived achievement, while a non-poor person will never become poor as a result of a decrease in the level of a deprived achievement' (Alkire and Foster, 2011a, p. 485).

The key value-added of a rigorously implemented MPI is that it conveys additional information not captured in single-dimensional measures, on the joint distribution of disadvantage and the composition of poverty among different multiply-deprived groups. 'Such an index also provides a consistent account of the overall change in multidimensional poverty across time and space as a supplement to single-dimensional measures, which should not be abandoned' (Alkire, 2011, p. 4)

The MPI includes more non-income indicators than the HDI. Further, while the HDI was intended to be used to measure the overall progress of a country's development, the MPI is concerned exclusively with a particular segment of the population, excluding information about the non-poor. The MPI identifies those who are poor through a two-step process involving identifying cut-offs of deprivation. 'The first is the traditional dimension-specific cut-off, which identifies whether a person is deprived with respect to that dimension. The second delineates how widely deprived a person must be in order to be considered poor' (Alkire and Foster, 2011a, p. 477). In this way, the MPI simultaneously concerns itself with *how many* people are experiencing poverty as well as *how much* (or the depth of) the deprivation is being experienced.

Two key properties mean that the methodology of the MPI can be seen to satisfy both Sen's view that poverty is a deprivation of capabilities and Atkinson's call for multidimensional indices to take full account of this complexity (Sen, 1993; Atkinson, 2003). These properties are that the MPI can be segmented into sub-groups (for example regional populations or ethnic groups) and that it can be sectioned to highlight which dimensions of poverty are most severe for the entire population (or for any sub-group).

The MPI approach also overcomes the weakness associated with the *union* method, which identifies a person as poor if they are deprived in one dimension of poverty, as well as the *intersection* method, which only identifies a person as poor if they are deprived across all the dimensions of poverty. The index also assists in targeting those poverty alleviation strategies that can address sub-populations, regions or specific deprivations (or combinations of these). In this way it provides a more nuanced assessment of deprivation, and thus poverty. However, this approach is not unique to the MPI and can be seen in other work (Mack and Lansley, 1985; Gordon, et al., 2003).

The MPI is calculated using the following formula:

$$MPI = H x A$$

where H is the headcount or the percentage of people who are identified as multidimensionally poor and A (intensity) is the percentage of dimensions in which the average poor person is deprived. A household is deemed poor if it is deprived in at least 33 per cent of the weighted indicators.

In multidimensional, as in single-dimensional poverty, H (the headcount) is familiar, intuitive and easy to communicate. It can be compared directly with an income poverty headcount, or with the incidence of deprivations in another indicator, and also compared across time. A (intensity) reflects the extent of simultaneous deprivations poor people experience. Table 5.2 below provides the dimensions, indicators, deprivation thresholds and weights for the MPI. Note that there is no direct indicator for income or consumption. This arguably makes the index particularly appropriate for measuring poverty in Melanesia given that the region lacks reliable data on income and that a large proportion of the region's population lives a semi-subsistence lifestyle.

Table 5.2 Dimensions, indicators, deprivation thresholds and weights for the Multidimensional Poverty Index

Dimension (weight)	Indicator (weight)	Deprived If...
Education (⅓)	Years of schooling (1/6)	No household member has completed 5 years of schooling
	School attendance (1/6)	At least one school-aged child is not attending school years 1 to 8
Health (⅓)	Child mortality (1/6)	A child has died within the house
	Nutrition (1/6)	Any adult or child for whom there is nutritional information is malnourished
Standard of Living (⅓)	Electricity (1/18)	The household does not have electricity
	Cooking fuel (1/18)	The household cooks on wood, dung or charcoal
	Floor (1/18)	The house's floor is dirt, sand or dung
	Sanitation (1/18)	The household does not have adequate sanitation (according to the MDG guidelines) or is shared
	Water (1/18)	The household does not have clean drinking water (according to the MDG guidelines) or is more than 30 minutes' walk away
	Assets (1/18)	The house hold does not own more than one of: radio, television, telephone, bicycle, motorbike or refrigerator, and does not own a car or truck

Source: Alkire (2011).

*The MP*I is not without its detractors. The arbitrary assignment of weights to its components is one criticism, as is the lack of an explicit linkage to conceptual analysis. It is argued that many multidimensional poverty indices – including the Physical Quality of Life Index (PQLI) (Morris, 1979), the HDI (UNDP 1990) and the MPI – are opaque, have hidden costs and downside risks that can lead to the distortion of poverty alleviation policies (Ravallion, 2010). Further, Ravallion argues that such indices are designed and based on the availability of data and for such composite indicators to have integrity, they must have conceptual clarity on what is being measured, transparency regarding any trade-offs within the index, and be able to survive robustness tests. In response MPI defenders link it to Sen's capability approach, emphasizing the transparency around its construction and demonstrating the index's robustness to a range of weights (Alkire, 2011; Alkire and Foster, 2011b).

5.2.4 Global Results

The MPI has been applied to 109 countries (accounting for 79 per cent of the global population). It shows that 31 per cent of the population (around 1.65 billion people) is poor. Half of all poor people live in South Asia, while 29 per cent live in Africa. An MPI for Vanuatu has been applied at the national level (though not for sub-regions). No MPI currently exists for Solomon Islands.

The majority of those deemed poor by the MPI reside in middle income countries – 1.19 billion compared to 459 million living in low-income countries. Interestingly, this reflects recent estimates of income-based poverty (Chandy and Gertz, 2011). The index indicates that whilst Africa's score is higher than that of South Asia, the poorest regions of South Asia have higher poverty rates and more people in poverty than sub-Saharan Africa. This sub-region analysis is useful as it shows, for example, that while Nepal has higher MPI-poverty than Cambodia, the poorest region of Cambodia is poorer than the poorest region of Nepal. Inter-temporal analysis indicates that poverty assessed by the index has reduced in Kenya, for example, through an improvement in the standard of living dimension, whilst Bangladesh's improvement resulted from reduced deprivations of all three dimensions of poverty.

5.3 Multidimensional Poverty in Melanesia

Although the OPHI has devised the MPI at the country level for Vanuatu, (relying on data from the UNICEF (2007) Multiple Indicators Cluster Survey (MICS) focusing on the health of children and women), data constraints have prevented the estimation of the MPI for the Solomon Islands. Moreover no regional index exists for these countries.

The household survey that was administered in these countries was designed to replicate the MPI for each of the 12 communities visited, as well as to collect specific information to tailor the index to the Melanesian context. A comparison of the MPI for other countries with the Solomon Islands and Vanuatu is provided. This is followed by a discussion of the tailored index for Melanesia: the MMPI.

5.3.1 The Multidimensional Poverty Index for the Solomon Islands and Vanuatu

The household survey in this study collected data on all of the key deprivations, with the exception of the malnutrition indicator. The nutrition deprivation cut off in the index is if any adult or child for whom there is nutritional information is malnourished (Alkire, 2011). Malnourishment is measured using anthropomorphic indicators. Adults are considered to be malnourished if their Body Mass Index (BMI) is below 18.5.

Collecting accurate measurements for an adult's height and weight (even if present) was not possible for the household surveyors. Instead a proxy must be used for whether there is a malnourished adult in the household. This proxy is best based on information regarding the food security situation of each household. As consistent access to adequate food for active healthy living is an important dimension of nutrition. Health survey questions from the US Food Security module (a self-reported indicator of behaviors, experiences and conditions related to food insecurity), were used in the household survey conducted in the Solomon Islands and Vanuatu. The US Food Security Module has been shown to be an inexpensive, easy to use analytical tool for evaluating food insecurity (Rafei, et al., 2009). Moreover, it has been successfully adapted for use in a wide variety of cultural and linguistic settings around the world – in particular in Asia and the Pacific (Derrickson et al., 2000).

Thus, as a proxy for malnutrition, responses to the following are used: 'Did you or any other adults in the house not eat food for an entire day because there wasn't enough money to buy food?' Food is generally the most pressing of priorities for any human being and to have gone without food for an entire day suggests severe food insecurity – particularly in the Pacific, where subsistence agriculture is so prevalent, social networks are strong and gift-giving is an ingrained cultural norm. Accordingly, if any household member is unable to draw upon these customary coping mechanisms for an entire day then the household's food insecurity situation is probably acute. Adults were the chosen as the appropriate referent object for food insecurity since the original index threshold asks whether there is 'any adult or child' that is malnourished. Given the tendency of parents to feed their children before feeding themselves, should children go without food for an entire day it clearly indicates more severe food insecurity (and one can doubtless infer that if a child in a household has gone without food, then so too have adults).

5.3.2 A Melanesian Multidimensional Poverty Index (MMPI)

Representing an important departure from the HDI, the MPI is a widely well-regarded measure of poverty that is able to be calculated for a large number of countries and can be modified to individual country contexts and priorities. This chapter tailors the MPI to include further information relevant to the nature of poverty in Melanesia. In modifying the index it is important that any indicators are objective and quantifiable, have clearly defined thresholds, can be categorized as a binary measure and, of course, are actually available.

In tailoring the MPI we introduce a new dimension of welfare – that of *access*. This dimension receives an equal weight to the other three dimensions of well-being.[1] Previous analysis illustrates that poverty in the Pacific is not about destitution, *per se*, but rather poverty of opportunity and a lack of access to key services (Abbott and Pollard, 2004). The importance of having access to a social support network is also a key aspect of well-being in Melanesia. Within the dimension of access we have devised three separate indicators of poverty: the produce garden, remoteness of services and the existence of a strong social network. Each of the four dimensions of wellbeing (health, education, standard of living and access) has been re-weighted to account for one quarter of the total weighting (compared with the one third that the three incumbent dimensions are each given in the standard MPI). The individual indicators for each of these respective dimensions have also been re-weighted accordingly (see Table 5.3). Each indicator and its deprivation cut-off are discussed in turn.

A garden is probably the most fundamental livelihood asset that households possess in Melanesia. Much of Melanesian culture revolves around the garden, both in terms of its fruits and the practice of gardening itself. Households that do not have access to a garden and its produce are therefore isolated from an important cultural activity and, more practically, must rely on the cash economy (or extended family favors) for their food. According to this indicator, a household is considered deprived if it reports not having access to a garden.

Remoteness of essential services is another important dimension of hardship in the Pacific, as identified by the ADB's Participatory Poverty assessments (Abbott and Pollard, 2004). The remoteness of many villages, and the funding constraints facing policymakers, results in a limited number of education and health providers. Additionally, access to centralized markets in which individuals can buy and sell a range of differentiated goods and services is also limited. This constrains the range of basic goods available for purchase and limits income earning opportunities. Specifically, a household is considered to be deprived if it takes more than half an hour to travel to a health service (hospital or clinic), a secondary school or to a market. While access to essential services might be partially picked up by other indicators of the index (in the health and education dimensions), this will not always be the case. The importance of access to services in Melanesia warrants the inclusion of a separate indicator.

Health is fundamental to human well-being and having good access to health clinics and hospitals is paramount during serious illness, injury or during child

1 The MPI and MMPI assign equal weights to the different dimensions of well-being. As noted above, Alkire and Foster (2011b) find that rankings of the MPI are robust to different weights. It is also true that the different dimensions of the MPI and MMPI are not highly correlated and they are therefore contributing additional information on well-being to the index.

Table 5.3 Dimensions, indicators, deprivation thresholds and weights for the Melanesian Multidimensional Poverty Index

Dimension (Weight)	Indicator (Weight)	Deprived if....
Health (¼)	Mortality (⅛)	Any child has died in the family
	Nutrition (⅛)	Any adult or child for whom there is nutritional information is malnourished*
Education (¼)	Years of schooling (⅛)	No household member has completed five years of schooling
	School attendance (⅛)	Any school-aged child is not attending school in years 1 to 8
Standard of Living (¼)	Electricity (1/24)	The household has no electricity
	Sanitation (1/24)	The household´s sanitation facility is not improved (according to the MDG guidelines), or it is improved but shared with other households
	Water (1/24)	The household does not have access to clean drinking water (according to the MDG guidelines) or clean water is more than 30 minutes walking from home.
	Floor (1/24)	The household has dirt, sand or dung floor
	Cooking fuel (1/24)	The household cooks with dung, wood or charcoal.
	Assets (1/24)	The household does not own more than one of: radio, TV, telephone, bike, motorbike or refrigerator, and does not own a car or truck
Access (¼)	Garden (1/12)	The household does not have access to a garden
	Services (1/12)	> 30 minutes travelled to health clinic or secondary school or market
	Social support (1/12)	Household has no one to rely upon in a time of financial difficulty

Note: A proxy measure was used for this indicator. A households is deprived if they answered in the affirmative to the question 'Did you or any other adults in the house not eat food for an entire day because there wasn't enough money to buy food' taken from the US Food Security module.
Source: Based on Alkire (2011).

birth. High rates of infant and maternal mortality in Melanesia reflect poor household access to these services and the resultant human suffering.

Access to a secondary (rather than primary) school is assessed for a number of reasons. Having no secondary school close by was a very common complaint made by focus group participants and key informants. Similar complaints were not registered against the proximity of primary schools – even in the most remote and rural areas. Moreover, primary school education is (notionally) free in Melanesia and enrolment rates are high. Consequently, remoteness from a primary school (if it exists) does not appear to be a major constraint on education. In contrast, secondary schools are much less widely available in the Solomon Islands and Vanuatu. Thus, when a secondary school is not nearby, families are often required to send their children to school as long-term boarders (AusAID 2012).

The financial costs and time spent getting to a main markets were major complaints of focus group participants. While small local markets exist in all communities, for example roadsides and at kava bars in Vanuatu, better income earning opportunities are available in central markets.

Strong social networks and the system of reciprocity are hallmarks of the traditional economy in Melanesia and key providers of a variety of important services (Regenvanu, 2009). Households that do not have anyone to rely upon in a time of need are therefore likely to be deprived of a key dimension of informal social security. Households are classified as deprived for this indicator if they are unable to rely on anybody in the event of someone in the household getting into financial difficulties and needing support. It is recognized that there is no objective measure of financial difficulty in this instance, and that the number of people relied upon is necessarily imprecise, but this information should nevertheless provide an indication of households that lie outside a social support network.

5.3.3 Analysis of the Multidimensional Poverty Indexes

Table 5.4 provides the incidence of poverty (H), the average intensity of poverty (A) and the index values for both the MPI and MMPI at a community and country level. Figures 5.1 and 5.2 also plot the incidence of poverty and the index value scores.

At a country level, according to the MPI, the Solomon Islands has a greater proportion of households that are deemed poor, relative to Vanuatu. Focusing first on the headcounts of MPI-poor in each country; according to the household survey data, one quarter of Solomon Islands households are MPI-poor. Relative to other developing countries this is figure is similar to Bhutan, Guatemala and Nicaragua. In Vanuatu an estimated 16 per cent of households are deemed MPI-poor, a rate similar to Tajikistan and Mongolia.[2] For the sake of comparison, in 2006, 26 per cent were estimated to be below the basic needs poverty line in the Solomon Islands and 16 per cent in Vanuatu (AusAID, 2009). The average intensity of

2 These figures rely on our household survey sample being nationally representative.

Table 5.4 Multidimensional poverty indices by location and country

Multidimensional Poverty Indices														
Headcounts of Poverty and Average Intensity of Poverty; Comparisons Across Location and Country														
	Honiara	Auki	GPPOL	Weather Coast	Malu	Vella	Vila	Luganville	Baravet	Mangalilu	Hog Harbour	Banks	Solomon Islands	Vanuatu
Multidimensional Poverty Index (MPI)														
Headcount ratio (H)	23.0	34.6	10.6	41.6	26.8	15.4	27.6	10.6	16.4	13.3	10.5	15.4	25.1	15.8
Average intensity (A)	41.4	41.6	44.4	46.9	42.4	38.9	43.8	43.8	43.9	42.8	38.2	38.9	43.0	42.3
MPI = H x A	0.095	0.144	0.047	0.195	0.114	0.060	0.121	0.046	0.072	0.057	0.040	0.060	0.108	0.067
Rank	5	2	10	1	4	7	3	11	6	9	12	7	1	2
Melanesian Multidimensional Poverty Index (MMPI)														
Headcount ratio (H)	20.7	41.0	11.8	49.4	19.5	19.2	34.5	12.9	19.4	18.7	7.9	11.5	26.5	17.7
Average intensity (A)	41.0	41.9	40.8	45.2	40.6	38.6	39.3	40.2	39.4	39.3	40.3	37.5	42.1	39.3
MMPI = H x A	0.085	0.172	0.048	0.223	0.079	0.074	0.136	0.052	0.076	0.073	0.032	0.043	0.112	0.070
Rank	20.7	41.0	11.8	49.4	19.5	19.2	34.5	12.9	19.4	18.7	7.9	11.5	26.5	17.7

Note: Sample size N=955

Source: The authors.

deprivation faced by the poor (*A*), is also relatively higher in the Solomon Islands. The survey data indicate that in the Solomon Islands, the average poor household is deprived on 43.0 per cent of the indicators compared with slightly less (42.3 per cent), in Vanuatu.

The MPI varies greatly across the communities surveyed in the Solomon Islands and Vanuatu and the data provide some important insights into the nature of poverty across location. With an incidence similar to Swaziland and the Republic of Congo, the remote Weather Coast (in the Solomon Islands) is by far the poorest location with 41.6 per cent of households deemed MPI-poor, followed by Auki with 34.6 per cent. The locations with the least incidence of poverty are the Guadalcanal Plains Palm Oil (GPPOL) villages and Vella Lavella with 11 and 15 per cent of households deemed poor respectively. In Vanuatu, the incidence of multidimensional poverty is highest in the capital Port Vila (with 28 per cent of households living in multidimensional poverty) and Baravet, Pentecost, with 16 per cent. The incidence is lowest in Luganville and Hog Harbour with about 11 per cent of households deemed MPI-poor.

In the vast majority of developing countries, poverty is predominantly a rural issue. However, findings from this household survey reveal that poverty in Melanesia is actually highest in urban areas (Port Vila and Auki) as well as in remote, rural areas (such as the Weather Coast in Solomon Islands and Baravet in Vanuatu). In contrast, the least-poor communities are those that are essentially rural in character, with good access to land and opportunities to earn income from agriculture and tourism and with effective transport links to market centres (such as Luganville, GPPOL villages and Hog Harbour).[3] Interestingly, when communities are aligned broadly in terms of their remoteness from main markets, a distinctive U-shape pattern emerges in the distribution of poverty (see Figures 5.1 and 5.2). While this may be partially the result of the sample consisting of squatter settlements in urban areas, it does highlight potential dangers of migrating to urban areas that have limited income-earning opportunities and limited access to land for productive gardening.

Combining the headcount rate of poverty with the average intensity of deprivations yields the MPI values for each region. At a national level the Solomon

3 It should be noted that Auki and Luganville are the second largest towns in the Solomon Islands and Vanuatu, respectively. The incidence of poverty in Auki tends to more closely resemble that of remote communities and capital cities while poverty in Luganville is more akin to the well-connected rural communities of GPPOL and Hog Harbour. In part, this may reflect the divergent economic fortunes of the two cities: in particular the steady stream of tourism to the east coast of Espirutu Santo that funnels through Luganville and is largely absent from Malaita. Indeed, it is likely to be no coincidence that Hog Harbour, which is connected to Luganville via the East Santo road, also performs relatively well on poverty and vulnerability metrics. This provides a cautionary tale of the importance of not over-generalising the results from 12 unique communities.

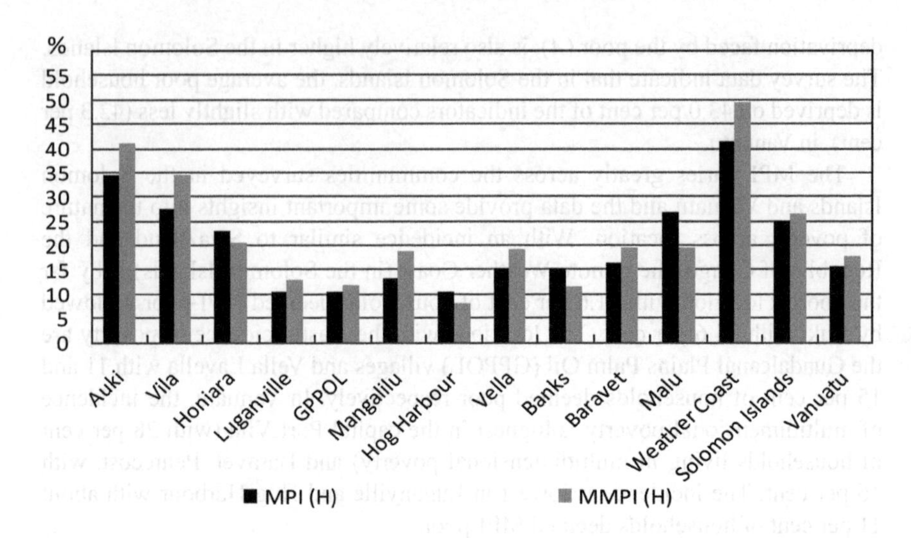

Figure 5.1 Percentage of households in multidimensional poverty by location and country

Source: The authors.

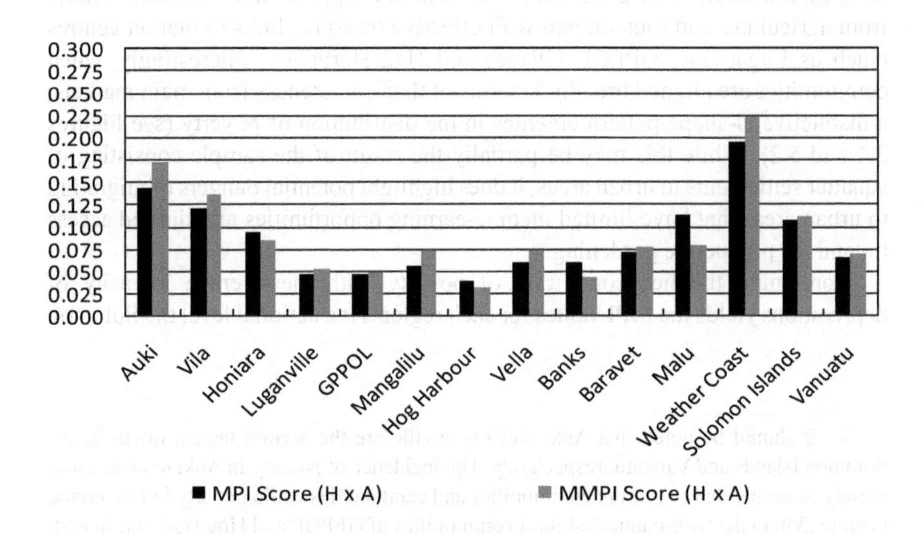

Figure 5.2 Multidimensional poverty indices (headcount x average intensity) by location and country

Source: The authors.

Islands has a MPI value similar to that of Lesotho, Sao Tome and Principe, and Burma. Vanuatu has a value comparable to Indonesia and Bhutan.

According to the MMPI, which incorporates data regarding various forms of access, there is a higher percentage of households that are poor across all locations except for Malu'u and Honiara in the Solomon Islands and Hog Harbour and the Banks Islands in Vanuatu. The higher incidence of MMPI poverty relative to MPI poverty in most communities is, in large part, due to the fact that most communities have a relatively high incidence of deprivation in the 'access' dimension. Urban regions stand out in terms of the lack of access to gardens, with 28 per cent of households, on average, across the four urban locations deprived in this indicator compared with only two per cent in rural communities. With the exceptions of the Banks Islands, Hog Harbour, Luganville and Honiara, communities recorded deprivation rates in the 'support' indicator in excess of 20 per cent, with Mangalilu and GPPOL recording deprivation rates in excess of 30 per cent. On the 'access to services' indicator, the geographically remote communities of Weather Coast, Vella Lavella and Baravet each recorded particularly high rates of deprivations, in excess of 80 per cent, reflecting a general lack of access to hospitals, secondary schools and markets (though access to a market in Baravet was much better than in the other two locations).[4] The provincial sub-station of Malu'u has the lowest rate of observed deprivation in the 'access to services' indicator, on account of the fact it is well serviced by a hospital, secondary school and market places. This evidence provides support for the importance of accounting for access when mapping the incidence and depth of poverty in Melanesian countries.

The MPI and MMPI can easily be broken down to examine how much each dimension contributes to multidimensional poverty. Figure 5.3 provides this information at the country level. The longer the bar, the greater the contribution of the dimension to overall poverty. The figure indicates that the standard of living dimension contributes most to the MPI, almost half of total poverty, in both the Solomon Islands and Vanuatu. This is followed by the health dimension and education. The health dimension makes a greater contribution to poverty in Vanuatu (33 per cent) than it does in the Solomon Islands (27 per cent). For the MMPI, the access dimension contributes an approximately equal proportion to poverty in both countries (around 27 per cent). Interestingly, access contributes more to poverty than each of the other dimensions apart from standard of living, highlighting the importance of tailoring poverty indices to country specific circumstances.

4 Somewhat surprisingly, the Banks Islands, a particularly remote community, the deprivation rate in the access to markets indicator was the lowest of all the communities surveyed. This probably illustrates one of the potential shortcomings of different perceptions of what a market constitutes. However, this is unlikely to substantially bias the results since the market component is but one of three indicators of services access (which, in turn only comprises one-twelfth of the MMPI) and the Banks had a relatively high proportion of households that were deprived according to the access to education indicator.

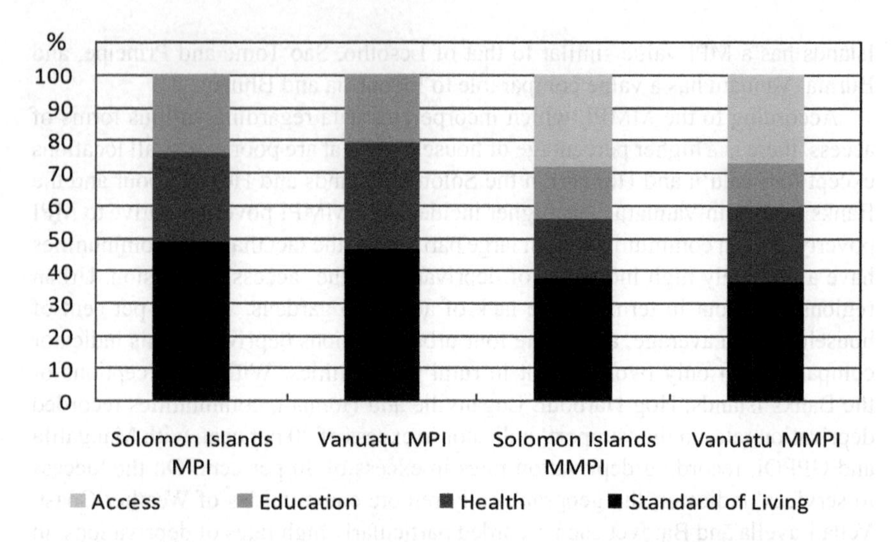

Figure 5.3 Percentage contribution of dimensions to multidimensional poverty by country

Source: The authors.

A further way of examining the indices is the identification of those who are severely poor and those that might be vulnerable to experiencing poverty. Disaggregating the indices can identify households that are severely poor (with weighted deprivations greater than 0.50 per cent and those that are less severely poor (those with a weighted deprivation between 0.30 and 0.50). Additionally, vulnerable households can be identified in the sense they fall just shy of the threshold value to be considered MPI-poor. Vulnerable households are those with a weighted average of deprivations somewhere between 0.20 and 0.30 (Alkire and Foster, 2011a). Results from this exercise are presented in Figure 5.4.

Using both the MPI and the MMPI, the share of households that are neither poor, nor vulnerable to poverty is much higher in Vanuatu than it is in the Solomon Islands. Using the MPI, 62 per cent of households in Vanuatu are not poor or vulnerable, compared with 47 per cent of Solomon Islander households. Using the MMPI, these proportions are 49 per cent and 32 per cent, respectively.

However, in both countries a large proportion of households are vulnerable to experiencing poverty: 23 per cent in Vanuatu and 28 per cent in the Solomon Islands according to the MPI, rising to 33 per cent and 42 per cent for Vanuatu and the Solomon Islands respectively for the MMPI. In fact, in each case a greater proportion of households are deemed vulnerable than are actually in poverty – considerably so in the case of the Melanesian index. Given the higher degree of exposure of Melanesian households to shocks, these vulnerable households face a high likelihood of experiencing poverty in the future.

A relatively small proportion of households are in severe multidimensional poverty. In Vanuatu 3.8 per cent of households have a weighted average of deprivations in excess of 50 per cent. This is almost half of the rate of severe poverty in the Solomon Islands (7.2 per cent). While only 2.1 per cent of households in Vanuatu are severely poor according to the MMPI, the rate remains at 7.2 in the Solomon Islands (though these are not the same households, with the correlation between the two measures of severe poverty in the Solomon Islands only 0.66. However these aggregate results mask some significant variations between the regions. In the Weather Coast only 2.6 per cent of households are neither vulnerable nor MPI-poor (and 2.7 per cent of households are neither vulnerable nor non-poor using the MMPI). The Weather Coast and Auki also have the highest rates of severe poverty, with 11.7 per cent and 11.5 per cent of all households severely MPI-poor, respectively.

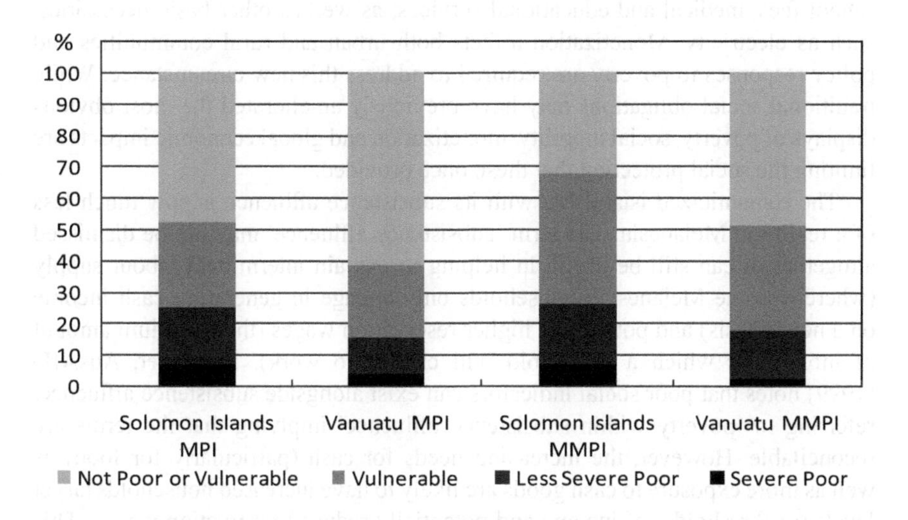

Figure 5.4 Multidimensional poverty, vulnerability and severe poverty by country

Source: The authors.

5.4 Conclusion

This chapter is the first to analyse multidimensional poverty at a regional level in Melanesian countries. Using unique household survey data conduced in 2010–11, it replicates the MPI from OPHI. It also tailors the index to better consider the Melanesian context, by including access to produce gardens, basic services and social support. The MMPI found that poverty in the Solomon Islands and Vanuatu not only varies between rural and urban locations but in general increased the incidence poverty reflecting the poor access Melanesian households often have to basic services. Multidimensional poverty was found to be highest in urban and remote locations.

For policymakers, the MMPI provides new insights into both the experience of poverty in the Solomon Islands and Vanuatu and the vulnerability to poverty. Central to poverty responses should be the recognition of the importance of family gardens for the production of staple foods. In urban locations, access to land to tend gardens is limited. Without this underlying means of self-support, Melanesians living in urban centers have an increased risk of being unable to meet their basic food needs. Recognition that monetization is a now an entrenched characteristic of Melanesian economies is also necessary for policymakers. Community life for families in the Solomon Islands and Vanuatu can no longer function effectively without access to a certain level of income. Cash is required to pay for basic necessities, including school fees, medical and educational services, as well as other basic necessities, such as electricity. Monetization affects both urban and rural communities and policy responses to poverty are required to address this new circumstance. While traditional social obligations may have previously ameliorated the most obvious displays of poverty, social mobility, monetization and global economic impacts are limiting the social protection that these once provided.

The romanticized island life with its subsistence affluence is now much less of a reality in Melanesia. The term 'subsistence affluence' may not be dismissed altogether. It can still be useful in helping to explain intermittent labour supply (whereby some Melanesian households only engage in generating cash income on a needs basis) and potentially higher reservation wages (the minimum amount of money for which a household will choose to work). Moreover, AusAID (1999) notes that poor social indicators can exist alongside subsistence affluence, referring to 'poverty within subsistence affluence' implying that the terms are reconcilable. However, the increasing needs for cash (particularly for food) as well as more exposure to cash goods are likely to have increased households target levels (or thresholds) of income and potentially reduced reservation wages. This is particularly true when there are increasing demands for cash at custom events.[5] Lifestyles are changing quickly and this chapter demonstrates that the harsh reality faced by a significant proportion of the Ni-Vanuatu and Solomon Islanders is a relatively high incidence of poverty and an even higher rate of vulnerability. Poverty is a real issue in the Solomon Islands and Vanuatu and cannot be ignored.

5 The issues are discussed in greater detail in the context of Papua New Guinea by AusAID (1999).

References

Abbott, D. and Pollard, S. (2004), *Hardship and Poverty in the Pacific: Strengthening Poverty Analysis and Strategies in the Pacific* (Asian Development Bank: Manila).

Alkire, S. (2011), Multidimensional Poverty and its Discontents, *OPHI Working Paper No. 46* (Oxford Poverty and Human Development Initiative, University of Oxford).

Alkire, S. and Foster, J. (2011a), Counting and Multidimensional Poverty Measurement. *Journal of Public Economics,* 95: 476–87.

Alkire, S. and Foster, J. (2011b), Understandings and Misunderstandings of Multidimensional Poverty Measurement, *Journal of Economic Inequality,* 9: 289–314.

Alkire, S. and Santos, M. (2010), Acute Multidimensional Poverty: A New Index for Developing Countries, *OPHI Working Paper, No. 38* (Oxford Poverty and Human Development Initiative, University of Oxford).

Atkinson, A. (2003), Multidimensional Deprivations, *Journal of Economic Inequality,* 1: 51–65.

AusAID (1999), *The Economy of Papua New Guinea: Macroeconomic Policies: Implications for Growth and Development in the Informal Sector*, International Development Issues No. 53, Prepared for AusAID by Economic Insights Pty Ltd, Australian Agency for International Development, Canberra.

AusAID (2009), *Tracking Development and Governance in the Pacific (*Australian Agency for International Development: Canberra).

AusAID (2012), Poverty, Vulnerability and Social Protection in the Pacific: The Role of Social Transfers (Australian Agency for International Development: Canberra).

Booth, C. (1887), The Inhabitants of Tower Hamlets (School Division Board), Their Condition and Occupations, *Journal of the Royal Statistical Society,* 50: 326–40.

Borguignon, F. and Chakravarty, S. (2003), The Measurement of Multidimensional Poverty. *Journal of Economic Inequality,* 1: 25–49.

Chandy, L. and Gertz, G. (2011), Poverty in Numbers: The Changing State of Global Poverty from 2005 to 2015, Brookings Global Views Policy Brief 2011-01 (The Brookings Institute: Washington).

Derrickson, J.P., Fisher, A.G. and Anderson, J.E.L. (2000), The Core US Household Food Security Survey Module Scale Measure is Valid and Reliable when Used with Asians and Pacific Islanders, *Journal of Nutrition*, 130: 2666–74

Doyal, L., and Gough, I. (1991), *A Theory of Need*, Macmillan, London.

Feeny, S. (2010), The Global Economic Crisis and the Millennium Development Goals in the Pacific, *Pacific Economic Bulletin*, 25(1): 136–50.

Feeny, S. and Clarke, M. (2008), Achieving the Millennium Development Goals in the Asia-Pacific Region: The Role of International Assistance, *Asia Pacific Viewpoint,* 49(2): 198–212.

Feeny, S. and Clarke, M. (2009), *The Millennium Development Goals and Beyond: International Assistance to the Asia-Pacific* (Palgrave-Macmillan: London).

Foster, J., Greer, J. and Thorbeke, E. (1984), A Class of Decomposable Poverty Measures, *Econometrica,* 52: 761–6.

Gordon, D., Nandy, S., Pantazis, C. and S. Pemberton (2003), *The Distribution of Child Poverty in the Developing World* (Centre for International Poverty Research: Bristol).

Kakwani, N. and Silber, J. (2008), Introduction: Multidimensional Poverty Analysis: Conceptual Issues, Empirical Illustrations and Policy Implications, *World Development,* 36(6): 987–91.

Mack, J. and Lansley, S. (1985), *Poor Britain* (Allen & Unwin: London).

MNCC (2012), Alternative Indicators of Well-being for Melanesia, Vanuatu Pilot Study Report (Malvatumauri National Council of Chiefs: Port Vila, Vanuatu).

Morris, D. (1979), *Measuring the Condition of the World's Poor: The Physical Quality of Life Index* (Pergamon: New York).

Narayan, D., Chambers, R., Shah, M.K. and Petesch, P. (2000), *Voices of the Poor: Crying Out for Change* (Oxford University Press for the World Bank: New York).

Nussbaum, M.C. (1988), Nature, Function and Capability, *Oxford Studies in Ancient Philosophy Supplement,* 1: 145–84.

Nussbaum, M.C. (1992), Human Functioning and Social Justice, *Political Theory,* 20(2): 202–46.

Nussbaum, M.C. (2000), *Women and Human Development: The Capabilities Approach* (Cambridge University Press: Cambridge).

PIFS (2011), *2011 Pacific Regional MDG Tracking Report*, Pacific Island Forum Secretariat, Suva.

Rafiei, M., Nord, M., Sadeghizadeh, A. and Entezari, M.H. (2009), Assessing the Internal Validity of a Household Survey-Based Food Security Measure Adapted for use in Iran, *Nutrition Journal,* 8(28): 1–11.

Ravallion, M. (2010), Mashup Indices of Development, Policy Research Working Paper No. 5432 (World Bank: Washington).

Regenvanu, R. (2009), *The Traditional Economy as the Source of Resilience in Melanesia* (Vanuatu Cultural Centre: Port Vila).

Rowntree, B.S. (1902), *Poverty: A Study of Town Life* (Macmillan: London).

Sen, A. (1984), *Resources, Values and Development* (Blackwell: Oxford).

Sen, A. (1993), Capability and Well-Being, in M. Nussbaum and A. Sen (eds), *The Quality of Life* (Clarendon Press, Oxford).

UNDP (1990), *The Human Development Report,* United Nations Development Program (Oxford University Press: Oxford).

UNICEF (2007), *Vanuatu Multiple Indicators Cluster Survey* (United Nations Children's Fund: New York).

Vandemoortele, J. (2009), The MDG Conundrum: Meeting the Targets Without Missing the Point, *Development Policy Review,* 27(4): 355–71.

Vandemoortele, J. (2011), A Fresh Look at the MDGs, *Journal of the Asia Pacific Economy,* 16(4): 520–28.

Chapter 6
Vulnerability, Resilience and Dynamism of the Custom Economy in Melanesia

Lachlan McDonald, Vijay Naidu and Manoranjan Mohanty

6.1 Introduction

Traditional economic systems have persisted in many countries of the global South despite its on-going incorporation into the global capitalist system. Typically, such systems are agrarian and static, with cultivation techniques little changed over long periods of time (Duffy, 2008). Polanyi articulated that the concept of the traditional economy is one where the economic decision making is fundamentally embedded within a broader socio-political framework; distinct from the market exchanges that characterize the modern free-market economy and the ideological prescriptions of the politburo in a command economy (Polanyi, 1944). To that end, allocative decisions within a traditional economy are subordinated to cultural or religious mores with rules, traditions and obligations guiding production, exchange and distribution. Examples of such traditional economic systems abound: Rosser et al. (1999) identify that in rural India a traditional economy, based in *jajmani* system associated with the Hindu caste system, has persisted in the villages and countryside despite the rapid encroachment of modern market forces and India's legacy of central planning. Rosser and Rosser (2005) identify that Islamic economics represents an important manifestation of a traditional economic system in which actors have concrete obligations to others and must conduct business in accordance with religious dictates, even at the expense of profits. Other notable religious and traditionalist views of society that have spawned non-market economic systems include Confucianism principles in East Asia, and Amish communities in North America (Rosser and Rosser, 2005).

In the Pacific, to varying degrees, all island states and territories continue to have traditional methods of production that exist alongside formal market systems. This is especially the case in the Melanesian countries of the Solomon Islands and Vanuatu, each of which is situated on the fringe of the global economy and has a high proportion of the population residing in rural communities, despite experiencing rapid urbanization over the past few decades. The custom (or *kastom* in the local languages of Pidgin and Bislama) economy relates to kinship group based production, distribution and consumption of horticultural and marine produce, and livestock, as well as other commodities of intrinsic value. Both foodstuffs and items of value are used in customary exchanges around life events and for dispute settlement. Similar to other traditional systems, in Melanesia the emphasis of the custom economy is a familistic groupism, that is, a system based on the *wantok* (a Pidgin English word meaning 'one who speaks the same language') system. While explicit *wantok*-ism is more prevalent in the Solomon Islands and Papua New Guinea than Vanuatu, it is a term that nonetheless resonates in Vanuatu, which has similar cultural practices (Nanau, 2011). According to Forsyth, this has been a stable form of social organization based on respect, reciprocity, shame, balance and the importance of the community with strong ties associated with lineage and kinship that predates the arrival of Christianity to the region (Forsyth, 2009).

To the extent that approximately 80 per cent of the inhabitants of the Solomon Islands and Vanuatu live in rural hamlets and villages, it is evident that the custom economy continues to play an important role in the livelihoods of Melanesians. The custom economy continues to provide sustenance, employment, socio-cultural relations, as well as cash income. It is therefore pivotal that consideration be given to it in any examination of change and development in these countries (Brookfield, 1973; Crocombe, 2001). This chapter shows that the custom economy can insulate households and communities from volatility in the national and global market economies and thus provides an important form of resilience. However, rapid urbanization and growing social inequality have placed pressure on the capacity of the custom economy to continue providing both livelihoods and informal social protection (Wood and Naidu, 2008; Naidu and Mohanty, 2009; Mohanty, 2011; Bedford, 2012). With land becoming increasingly scarce, particularly in urban areas, and incomes falling in real terms, meeting customary obligations and assisting members of the extended family has become a burden for some households, and even jettisoned by some as they've sought to cope with recent economic shocks.

Using a unique household survey, this chapter explores the Melanesian custom economy and the pressures that have been brought to bear on it in the face of structural shifts in the economies of the Solomon Islands and Vanuatu. Section 6.2 outlines the antecedents of the custom economy, Section 6.3 details the contemporary custom economy while Section 6.4 summarizes the results from the household survey that examines the state of the custom economy and its capacity to provide resilience in the Solomon Islands and Vanuatu. Section 6.5 concludes with policy options.

6.2 The History of the Custom Economy in Melanesia

The antecedents of the Melanesian custom economy stretch back millennia. There is evidence that over 3,000 years ago inter-island trade existed over an area of around 1,500 square miles, with obsidian and perhaps lapita pottery being exchanged for unknown items (Shutler and Shutler, 1975). Over time, trading circuits evolved between different islands within the region on the basis of the different endowments of natural resources and skills that each location possessed

(Belshaw, 1965; Hughes, 1978). Coastal communities traded sea-derived commodities in return for the natural resources from inland areas. Poorly resource-endowed islands specialized in manufacturing, including mats, shell ornaments, wooden bowls and canoes, and traded these for foodstuffs and raw materials.

The Kula Ring is the best known of these trade circuits, involving Papuan coastal and neighbouring islands' communities. Exchange, presentations, gifts and commodities operated very much within culturally defined ways and often were integral to the social relationships between the groups involved (see Douglas, 1967).

Rather than impersonal market forces, persons of rank, including chiefs and 'big men', have historically played central roles in the process of production, accumulation and distribution in Melanesia. According to Narokobi, in traditional Melanesian societies, 'leaders by definition are distributors of wealth. Custom requires them to share their personal wealth' (Narokobi, 1983, p. 10). 'Big men' thus coordinated and mobilized labour so that there was a surplus of food, luxury items and valued goods with the express purpose of engaging in exchange relations that helped maintain and reinforce social relations. A number of life-cycle events then triggered processes of exchange. These included deaths and mortuary feasts, bride price and marriages, as well as disputes and conflicts. A sophisticated trading regime supported these activities, with goods of similar practical and cultural value exchanged. Foods for subsistence were exchanged with each other, whereas luxury items and items of wealth were traded separately (Sillitoe, 2006). Chowning notes that 'in all Melanesian societies, wealth consists primarily of two things: domestic pigs and some sort of small portable valuables, usually made of seashell. There are always additional valuables, but the primary ones are considered essential to major transactions, such as marriage payments and compensation for death' (Chowning, 1977, p. 49). In addition to bestowing wealth, pig killing ceremonies are also important methods held to resolve disputes and conflicts, and celebrate momentous occasions (Lamont, 1999).

While it is likely to have been the case that potlatch and feasts resulted in the destruction and distribution of items of wealth, cultural transactions served an important role in constraining the extent of economic inequality, and conferring social standing and rank to both chiefs and 'big men' (Sillitoe, 2006).

6.3 The Custom Economy in Modern Melanesia

Just like the system of land tenure that forms its foundation, the custom economy in Melanesia is complex and diverse (Crocombe, 2001). In the context of the law, Forsyth notes that perhaps it is unreasonable to consider the *kastom* system in Vanuatu a 'system' at all, given the considerable heterogeneity of language groupings and indigenous legal systems. However, she notes that 'there are common threads that unify them: the emphasis on peace and harmony in the community, on restoring relationships, on the use of chiefs to facilitate agreement, on community involvement in the processes and on the achievement of settlement by the payment of compensation' (Forsyth, 2009, pp. 95). Similarly, in the economic system, while referring to a monolithic 'custom economy' risks being overly reductive, sufficient broad characteristics of the Melanesian custom economy are nonetheless discernible so as to permit some generalizations. In particular, in line with most traditional economic systems, the custom economy is characteristically based on mixed farming. Land provides a critical sense of socio-cultural identity in Melanesia, and is a key binding mechanism of families, clans and tribes. As Ratuva (2006, p. 103) points out, 'land is inalienable from social life and vice-versa'.

Ni-Vanuatu Parliamentarian Ralph Regenvanu groups his native Vanuatu together with Melanesian neighbours Solomon Islands and Papua New Guinea when discussing the importance of land in the Melanesian custom economy (which he refers to as 'the traditional economy'). He stresses that a central benefit of the system is that it provides 'excellent sustainable management of the natural environment' (Regenvanu, 2009, p. 31). This rich natural capital, he argues, provides relatively equitable access to horticultural plots, which gives the broad masses of Ni-Vanuatu food security, shelter, medicinal plants and other necessities of life. Gardens produce root crops, such as sweet potatoes, taro, yams, and green vegetables as well as coconuts and bananas. Natural building materials, such as sago palm and pandanus leaves, are also cultivated. Haccius (2009) notes that the practicalities of shifting cultivation techniques of such subsistence agriculture requires a flexible system of land. Consequently, ownership of contiguous land is blurred and usufruct rights are extended to members of the extended family or sub-clan, based on language grouping, lineage, clans and tribes with common ancestors, totems and cultural practices. The fruits of the land are then used as consideration in these agreements. The Malvatumauri National Council of Chiefs (MNCC), the peak body in charge of custom and tradition in Vanuatu, notes that land is a public good, and 'the argument is often made that no one "owns" land in Vanuatu and that families and the individuals within a family unit are better described as custodians of the land' (MNCC, 2012, p. 20).

The importance of the land in Melanesia is reflected in the fact that, in Vanuatu at least, custom land holding remains the foundation of land ownership and is enshrined in the Constitution. Bolton notes that in the pre-independence meetings of the committee that devised the Vanuatu Constitution, Walter Lini, then leader of the Vanuaaku Pati, said, 'Land is the root of kastom. To deny customary owners their land would be to deny kastom' (Bolton, 2003). Consequently, it is notionally prohibited to sell land in Vanuatu; rather it must be converted to long term leases (over a maximum 75 year period). However, to the extent that leasing of land is commonly referred to as 'selling' probably reveals the view of many ni-Vanuatu that the land will not be returned (Haccius, 2009). In contrast, customary land may be leased or sold in the Solomon Islands, involving the Commissioner of Lands acquiring the land (through lease or purchase) after a public acquisition hearing. Registration of customary land is otherwise optional and has been little practiced, though more than 80 per cent of land is held in accordance with current customary usage (Monson, 2010).

Indeed, despite rapid urbanization, the custom economy remains central to Melanesian livelihoods. Some 80 per cent of the populations of the Solomon Islands and Vanuatu currently reside in rural areas and domestic production continues to provide a substantial share of the total food consumed in each country. In addition, apart from the poorest and those in living in crowded urban settlements, urban households also rely on the garden for food (Chung and Hill, 2002; Maebuta and Maebuta, 2009; UNICEF, 2011). McGregor et al. (2009, p. 26) assert that smallholder subsistence farming systems have represented a 'hidden strength of otherwise structurally weak economies' in Melanesia given their closed-loop nature of production and their capacity to provide access to nutritional food.

Jolly (1994) provides an account from South Pentecost in Vanuatu of the environment providing the context for the customary division of labour between men and women in Melanesia. Men commonly undertake the more physically demanding tasks including clearing land of bushes and trees and preparing it for cultivation. The building of houses and other shelters from locally grown materials, the slaughter of animals, the making of canoes and weapons are also men's responsibility. Hunting and deep-sea fishing in outrigger canoes, using both traditional methods and imported equipment that have been made and acquired by men, are also considered exclusively the domain of men. Women tend to predominate in domestic duties around the home, involving the caring for children and the invalid, as well as being responsible for the daily feeding and care of domestic herds of pigs kept for rituals. However, ownership of the pigs nonetheless resides with men, and their exchange and sale during traditional rituals is typically exclusively the domain of men (Bolton, 2003). Women, in contrast, are responsible for weaving local plants into traditional textiles used in these rituals (Jolly, 1994).

Jolly notes that the overall sexual division of labour is significantly influenced by the fact that 'men and women have a significantly different toolkit' to employ for such tasks (Jolly, 1994, p. 74). However, Jolly also observes that these gendered differences are due to the fact that men dominate in decisions on how household labour is organized. However, in terms of the practical work done in the in the food garden, Jolly also emphasizes that 'there is no rigid sexual division of labour', with men and women performing somewhat interchangeable roles (Jolly, 1994, pp. 63–4). Indeed, while men did the heavy work of clearing, burning and harvesting, women were responsible for weeding, harvesting, collection of water and firewood, and carrying of produce.

The natural environment also represents a critical intersection of the custom and modern economies. Local systems of exchange of marine and horticulture resources also provide many households with a critical source of cash. Surplus crops are directly traded for cash balances: copra, cocoa and coffee function like ordinary cash crops, while commodities such as betel nut and kava, which were formerly items of prestige and custom, have since been commoditized and are now widely peddled. Households also sell surplus produce to fund everyday expenditures and to fund investment in human capital and buildings (Gibson and Nero, 2008). In Vanuatu this is confirmed in the recent Agricultural Census, with 60 per cent of rural households nationwide selling one-quarter of their harvest, and a further 20 per cent selling half of the harvest (GOV, 2007).

Another defining characteristic of the custom economy in Melanesia is its redistributive function and its important social protection properties. Ward notes that '[Melanesian] society is not individualized to the extent that people are left destitute. Reciprocity still operates; kinship and place of origin groups support those in need, and income is spread through traditional-type channels' (Ward, 1977, p. 43). This was a theme that Regenvanu picked up more than three decades later when he espoused that 'there is a sense of a shared idenity, "community" and "belongingness" among the large extended family groups that make up the basic building blocks of Vanuatu society. This gives a high level of social security for all family members' (Regenvanu, 2009, p. 31). Manifest in various guises, such as contributions at death ceremonies, and sharing food and resources at community fundraising events, this strong sense of shared identity ensures that social insurance is available for kinfolk when needed.

The reciprocal flow of numerous goods, cash and services, as well as gift-giving on special feast days and cultural norms of social obligation, thus provide households in the Solomon Islands and Vanuatu with a social safety net that is largely absent in a formal sense (Ratuva, 2006). Mortuaries, marriages and misconduct represent important triggers for the system. Such informal social protection systems, in which risk is pooled amongst a community, are a common feature of developing countries where formal insurance markets, or public safety nets are unavailable (Alderman and Paxon, 1994; Dercon, 2002).

The mutual insurance provided by the custom economy has been variously nominated as a key reason why hunger and absolute destitution are not prominent in Melanesia (Abbott and Pollard, 2004) and also why households in the Solomon Islands and Vanuatu tend to be relatively insulated from global economic shocks. Regenvanu argues that the system 'constitutes the political, economic and social foundation of contemporary Vanuatu [sic] society and is the source of resilience for our populations, which has allowed them to weather the vagaries of the global economy over past decades' (Regenvanu, 2009, 30–31). Feeny (2010) suggests that the system was also critical in acting as a buffer to the transmission of externally-generated macroeconomic shocks associated with the food and fuel price crises leading into the Global Economic Crisis (GEC) and the GEC itself.

While custom relations are an integral part of modern Melanesia, non-monetized exchange and traditional methods of production and distribution are often excluded from more formal analyses of progress, such as GDP or GNI, and even measures of well-being, such as the Human Development Index (HDI). Consequently, the true level of Melanesian welfare is likely to be understated by these measures. In response to these perceived deficiencies, the MNCC in Vanuatu, with the support of the main regional bloc: the Melanesian Spearhead Group (MSG) has sought to enumerate alternative measures of well-being in Melanesia. At the 2008 MSG Leaders' Summit, leaders agreed that MSG governments 'should be better able to account for and measure the substantial non-cash values that contribute to their people's quality of life'. Consequently, they have reported on a series of alternative indicators of well-being that reflect Melanesian values. Unsurprisingly, the custom economy is front and centre in this pilot study, with variables grouped into three broad domains, including: resource access (including land and marine resources); cultural practices (including language, items of traditional wealth and ceremonial practice); and community vitality (MNCC, 2012).

However, a number of scholars have suggested that the strength of the traditional economy as a social safety net may be somewhat overplayed. Monsell-Davis argues that while gift-giving is undoubtedly central to Melanesian culture, the 'popular romantic' view of extended family ensuring that anyone can migrate home to the village, or receive food and shelter from urban kin jars with the empirical reality that pockets of destitution exist (Monsell-Davis, 1993). Moreover, he notes that expectations of reciprocity as well as the social stigma attached to being seen as refuting a request for help is actually the lens through which modern *wantok*-ism is viewed. He argues, therefore, that cultural obligations of distribution can be financially burdensome for households with limited resources, and that the relationship of exchange between households was often based on the expectation of receiving something in return. He cites evidence from Fiji that giving was more likely to be horizontal, rather than vertical, and therefore based less on genuinely redistributive principles but rather more akin to a loan (Barr, 1990, cited in Monsell-Davis, 1993).

A number of authors have also observed that the various exchanges and social obligations have become overly onerous (Lamont, 1999; Bre, 2006; Sillitoe, 2000). Cash is now very much a part of traditional exchanges as are a range of durable goods associated with the wider capitalist economy. New social obligations that include giving donations and tithes to churches and looking after *wantoks* outside the immediate family in urban areas add to the social and economic demands on urban families.

The shifting burden of the custom economy is also being manifest through the increasingly high prices being exacted in traditional exchanges – especially those related to marriage. As noted earlier, in addition to traditional valuables such as pigs and seashell, bride price exchange includes cash and even modern durable goods (which cannot be carved up and shared). Lamont (1999, p. 202) observed that in urban Papua New Guinea women of marriageable age are referred to as 'Toyotas' because of the rather large increases in the monetary value of bride price demanded by the parents of girls.

In addition to the general commodification of traditional exchanges, rapid urbanization in Melanesia also presents a particularly acute challenge to the custom economy. Rates of urbanization in the Solomon Islands and Vanuatu are amongst the most rapid in the world. While such growth is coming from a particularly low base, the urban population as a percentage of the total is nonetheless expanding rapidly, having tripled in each country since the 1960s (World Bank, 2012). Connell attributes the rapid urban drift to a combination of modern transportation, as well as stagnation of rural development and increasing significance of urban economies in global markets, which has increased both the 'opportunity and logic' for migration (Connell, 2011, p. 123).

With inadequate planning strategies to manage urban growth, inward migrants are tending to settle amongst ethnic brethren in squatter communities. These are typically characterized by unplanned, high density living in low-cost temporary housing and are often situated on marginal land on the fringes of town with limited provision of essential services, such as water and electricity (Chung and Hill, 2002; Maebuta and Maebuta, 2009; Kidd et al., 2010).[1]

In their survey of urban squatter communities in Vanuatu, Chung and Hill (2002) found that informal settlements were diverse communities: while some were relatively more secure in their land tenure, with permanent dwellings and a food garden, many in the most crowded settlements did not have this security. Most, however, were materially poor and faced difficulties in meeting their basic needs for food, clothing and money, as well as cultural obligations, because of a combination of reliance on wage income and insufficient labour demand (Chung and Hill, 2002). UNICEF (2011) discovered that squatter households without a garden were therefore acutely exposed to rises in food prices, as they had few alternatives for sourcing food. Critically, the lack of a garden also means that those households are alienated from a central element of the custom economy. Maebuta and Maebuta (2009) found similar characteristics in the squatter settlements of Honiara, and discovered that while households generally had access to wage income, often this was limited to one individual who was responsible for a large number of dependents. Indeed, it is for this reason that McGregor et al. (2009) find poor urban households in Melanesia were particularly vulnerable to the effects of the recent rises in international food prices.

According to Connell, the upshot of these cultural and structural shifts in Melanesia is that 'there is now good evidence that [rural and urban safety nets] are breaking down and it is no longer possible, if it ever were, for urban people simply to return and be supported by rural kin, while urban households are increasingly reluctant to host impecunious and unproductive rural kin' (Connell, 2011, p. 126). He also suggests that low incomes and a lack of support in squatter settlements have meant that some traditional obligations have been abandoned by some households simply to get by. For instance, cooperative labour has been replaced with individualism, materialism and competitiveness (Connell, 2010). Waring

1 In Vanuatu, Chung and Hill (2002, p. 10) note that the fastest growing regions of urban areas are the informal settlements, which have grown at twice the rate of urban population growth in general, while in Solomon Islands Maebuta and Maebuta (2009, p. 119) note that unauthorised settlements were growing at 26 per cent per annum in 2006 and that of the total population of Honiara (50,000), some 17,000 were illegal settlers living on government land.

and Sumeo substantiate this, to some extent, in their analysis of informal labour market responses in the Pacific more broadly to the recent economic crises. The authors cite the fact that aspirations for modern goods have led to a weakening in traditional family systems and raised expectations of monetary rewards for 'work', including caring work, even in rural areas. This, in turn, has changed attitudes toward social protection, with growing expectations that the government, rather than the extended family unit, should be responsible for providing caring services (Waring and Sumeo, 2010).

6.4 The Custom Economy in the Solomon Islands and Vanuatu

This rest of this chapter seeks to assess the current state of the custom economy in the Solomon Islands and Vanuatu. In the context of the recent international macroeconomic shocks associated with the food, fuel and economic crises, it sheds light on the extent to which the custom economy provides households with resilience, as well as examining whether elements of the custom economy are, indeed, proving overly onerous on households. Although there are regular rounds of Household Income and Expenditure Surveys in the Solomon Islands and Vanuatu, as well as occasional demographic studies (such as the Multiple Indicator Cluster Survey in Vanuatu and the Demographic Health Survey in the Solomon Islands) there is nothing in these contexts that captures the full gamut of information on demographics, household assets, income and expenditure. This research addresses this gap, through the use of a unique household survey that captures each of these dimensions as well as capturing information on cultural practices and important non-monetized production and exchange. In particular, the survey permits the examination of a range of important dimension of the custom economy, including: the role of the extended family within the household; security over land tenure; the availability of garden resources; the role of custom money; and the dynamics of the *wantok* system, including its role as a source of resilience from shocks. This chapter also draws on information from focus groups that were held separately with men and women.

In early 2011, more than 1,000 households were surveyed in the Solomon Islands and Vanuatu across 12 communities that were targeted to reflect the different livelihoods and different levels of engagement with both the modern and custom economies. In the Solomon Islands this included communities in Honiara (specifically in the squatter communities of White River and Burns Creek), Auki (in particular Lilisiana and Ambu), villages servicing Guadalcanal Plains Palm Oil Limited (GPPOL), the Weather Coast in south east Guadalcanal, Malu'u in north Malaita and remote communities on the south east coast of the island of Vella Lavella in Western Province. In Vanuatu, communities that participated in the research were from Port Vila (specifically in the squatter communities of Blacksands and Ohlen), Luganville (in particular Sarakata and Pepsi), Mangalilu, Hog Harbour in Santo, Baravet in south Pentecost and communities on the island of Mota Lava in the Banks Islands. For the purposes of analysis, communities were split into 'urban' and 'rural' (or, more-accurately, non-urban) with the former comprising the communities in Honiara and Auki in the Solomon Islands and Port Vila and Luganville in Vanuatu, respectively.

6.4.1 The Importance of the Extended Family

The importance of extended family in Melanesian countries is captured by the findings from the household survey. It reveals that 22 per cent of households have a child that is from their non-immediate family and about 9 per cent of households have a child from their extended family while not having a child of their own. Adoption arrangements, and the care of children of relatives in the kinship network or *wantoks*, are important features of social protection mechanisms in the custom economy. Orathinkal and Vansteenwegen (2004) note that in the rural areas, the social and family system is complex, with extended family and kinship systems. In contrast, they note that nuclear families are on the rise in the urban settings. While the stylized rural dwelling in Melanesia has a number of generations residing in the same familial dwelling, including grandparents, parents (and paternal kin if patriarchal, and maternal kin if matriarchal) as well as married and unmarried children, the survey, instead, suggests that urban households are larger than rural dwellings (5.9 people compared with 5.1 people) and have a greater percentage of non-immediate family members (15 per cent versus 10 per cent). However, to the extent that urban households are, on average, younger (34.9 years) than rural households (37.4 years) and have a higher proportion of males, then this may be capturing the influence of internal migration into urban areas.

6.4.2 Security of Land Tenure

Given the importance of land to Melanesian identities and livelihoods households were asked about the security of their tenure over the land that they reside upon. One way to look at the security of land tenure is to examine the proportion of households that pay any rent on their dwelling. Paying rent to live on a parcel of land is likely to suggest that a household is deprived of control over the land. Empirically, renting status has also been found to be associated with poverty and vulnerability in the Pacific, owing to the insecure living status and conditions of renters, and the fact that they have limited legal protections (Chung and Hill, 2002; Storey, 2006). On this score, only 5 per cent of households surveyed pay rent, though there is a wide variation across rural and urban areas, with around 13 per cent of households in urban squatter settlement areas regularly paying rent, compared with less than 2 per cent in rural areas. The relatively higher proportion of renters in urban regions is likely to reflect the increasing fragmentation and shortage of land in urban and peri-urban localities.

6.4.3 The Garden

It is well acknowledged that the garden, where traditional root crops are cultivated and harvested, is central to the livelihoods of most Melanesian households. In addition to being an important food source, horticultural resources are also an important source of income with surplus food sold at local marketplaces. However, few households (if any) are likely to derive their complete subsistence from the garden: rather, most households blend semi-subsistence from their garden with consumption of store-bought food (Jansen et al., 2006).

The survey indicates that about 88 per cent of all households nominated that they have access to a garden. Disaggregating these data into individual communities, it is clear that urban communities as a group have relatively less access to a garden, with 71 per cent indicating access, compared with the near-universal access in rural areas (96 per cent). However, even the rural–urban divide masks some important differences between communities. As demonstrated by Figure 6.1, some squatter settlement communities in urban areas are substantially more limited in their access to cultivable land. For instance, in both White River, in Honiara, and Lilisiana, in Auki, only around 43 per cent of households reported having a food garden. In Ambu, also in Auki, the rate was somewhat higher at 67 per cent, though still well below the levels elsewhere. Similarly, in Vanuatu the lowest rates of garden access were in the crowded urban areas of Ohlen in Port Vila and Sarakata in Luganville.

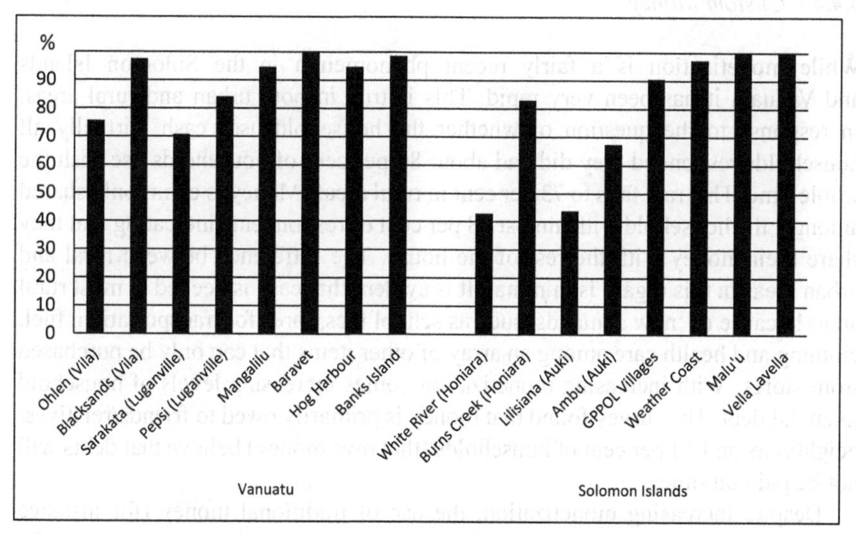

Figure 6.1 Percentage of households with a food garden by community

Source: The authors.

The relative limitations on gardening land in some urban communities belie an important fact about contemporary Melanesia: that access to indigenous food is not universal. Indeed, only 60 per cent of the total sample indicated that the garden was their household's primary source of food, with the rate falling to 31 per cent in urban areas. In Auki, this figure drops to an alarming 6 per cent. A further survey question asked households to indicate whether accessing food had become easier or more difficult in the two years preceding the survey. Urban communities were almost twice as likely to indicate that accessing food had become more difficult than rural households with 42 per cent indicating that was the case. Given their difficulties accessing food, it is perhaps unsurprising that residents of Auki were the most pessimistic of the communities surveyed, with 60 per cent indicating that conditions had deteriorated. The stress facing households with limited accessibility to gardens is also being manifest in food insecurity: 15 per cent of the total sample indicated that they'd experienced a day in which adults had to go without food for entire day. That this is occurring in Melanesia, where food is considered to be plentiful is alarming in itself, though the fact that this rises to 23 per cent in urban regions (and 30 per cent in Auki) should raise alarm bells amongst policymakers.

While the shortage of accessible arable land in urban areas is a constraint on gardening, it is noteworthy that the produce brought into urban households by visiting rural relatives form critical parts of their food source. In fact with the very low wage income, the reliance on these food sources helps in the survival of urban families (discussed below).

6.4.4 Custom Money

While monetization is a fairly recent phenomenon in the Solomon Islands and Vanuatu it has been very rapid. This is true in both urban and rural areas. In response to the question of whether the household uses cash, virtually all households responded they did and about 80 per cent of households use cash the whole time. This rate falls to 73 per cent in rural areas. Money is commonly shared amongst the household with almost 88 per cent of respondents indicating that they share their money with the rest of the house. The difference between rural and urban areas in this regard is minimal. It is evident that cash is needed in most rural areas because of 'new demands' such as school fees, fares for transportation, fuel, clothing and health care among an array of other items that can only be purchased from stores. With increasing monetization comes increasing levels of household financial debt. The survey found that money is primarily owed to friends/relatives/neighbours and 14 per cent of households (that owe money) believe that debts will not be paid on time.

Despite increasing monetization, the use of traditional money (for instance shell money, mats, and pig tusks and dolphins' teeth) is still common in the Solomon Islands and Vanuatu. Traditional money has important intangible benefits, linking communities and cultural areas and, as such, is highly prized. Indeed, the importance of traditional money in the cultural practices of Melanesia was highlighted in the Vanuatu Cultural Centre's '2007 – Year of the Traditional Economy', and UNESCO launched a project to examine the salient characteristics of traditional money and banking. The author of that report found that, similar to modern money, traditional money has medium of exchange and store of wealth properties and is still used in various rituals, including marriage, initiation ceremonies, and dispute resolution ceremonies (Huffman, 2005). He even discovered that 'economic concepts such as loan, credit, investment, interest and compound do exist in traditional cultures … however, the depository or "bank" is not a social institution as such but rather individuals and relationships' (Huffman, 2005, p. 12). Additionally, it was discovered that school fees in remote areas are sometimes paid with mats and pigs and afforded children that would ordinarily have been unable to attend the school the chance to do so.

According to the survey, custom money is used widely in both countries, though it appears to be more prevalent in the Solomon Islands, with 75 per cent of households responding that they use custom money at least some of the time, compared to 39 per cent in Vanuatu. Focus group analysis indicated that the use of 'red money' (that is custom money in the forms of strings of shells) was an important part of the price paid for a bride in the Solomon Islands. While this was not explicitly mentioned in Vanuatu, another way of looking at the data in Vanuatu is to examine the revealed preferences of households for traditional money by looking at the different types of gifts that households give at various ceremonies. On this measure, the percentage of households in Vanuatu that gave a woven mat at a wedding in Vanuatu rises to 62 per cent. This study provides just a snapshot of some dimensions of traditional money and traditional monetary systems are an interesting area for further research.

6.4.5 *Custom Events*

The survey also asked households questions regarding the types of gifts that are given at different ceremonies such as weddings, land disputes, funerals. The survey found that food and cash contributions are more commonly provided at a wedding in the Solomon Islands and Vanuatu relative to other gifts (such as livestock and woven mats). Cash is also likely to be given at land disputes, funerals and fundraising events while food is also commonly given at funerals. It is likely that cash and food are both preferred and more convenient items of exchange compared to pigs and cattle, and traditional artefacts such as mats because the latter require years of nurturing, and also make more intense demands on labour time and skills. Overall these findings suggest that while the traditional practice of giving non-cash gifts at ceremonies remains prevalent, cash is also a preferred gift medium.

Households were also asked about whether they have changed their gift-giving practices in recent years. A substantial proportion of households, 47 per cent, indicated that they give less at weddings, land disputes and funeral ceremonies than they did two years prior to the survey. This proportion was much higher in Vanuatu, where 60 per cent of households had decreased their offerings, compared with 34 per cent in the Solomon Islands. That a substantial percentage of households surveyed are donating less at these events is instructive and may provide some *prima facie* evidence of households reallocating their expenditure patterns in light of the increasingly expensive cost of living.

6.4.6 Support from Family and Wantok

Support from family is commonplace in the traditional social protection system that prevails in Melanesia. A majority of households surveyed, nearly 83 per cent, reported that they rely on family and *wantok* at least somewhat, with just 17 per cent of households indicating that they were not reliant on family and *wantok* at all. That the rates of support are high in both Vanuatu and the Solomon Islands, with 89 per cent and 77 per cent respectively, is unsurprising. Perhaps it is more surprising that so many households indicated that they did not rely on these social resources. As a cross-check to see whether there were indeed households that felt as though they were on their own, the survey also asked respondents how many people outside their household they could rely upon if someone in the household got into financial difficulties. A full 22 per cent replied that there was no one; with the result as high in rural regions as urban.

To examine the strength of the traditional social insurance system further, the survey asked about the contributions people make in cash or in kind towards the meeting the needs of other households (including food, school fees and cash remittances). The survey found that 44 per cent of households spend money on purchasing someone else's food each month, nearly 19 per cent of households spend money on other children's school fees and 31 per cent of households send cash to others each month (37 per cent households in urban areas compared to 28 per cent in rural areas).

Additionally, the survey examined the extent to which goods and money flow in both directions. The strength of the custom economy, rests, to a certain extent, on norms of reciprocity, in which important good and services flow in numerous directions (Regenvanu, 2009). Indeed, this is evident in the survey, with a substantial share of households indicating that they provide largesse to and receive it from family and *wantok* networks, in the form of: cash (66 per cent); clothes (58 per cent); and food (83 per cent). These figures provide evidence that social support systems within both countries remain robust, though gaps clearly exist.

6.4.7 The Custom Economy as a Source of Resilience

The survey also examined the different ways households in the Solomon Islands and Vanuatu responded to an unexpected negative event (or 'shock'). Results indicate that households in the Solomon Islands and Vanuatu have a very high degree of exposure to a range of different shocks. These include events that simultaneously affect all households within a given location, such as a natural disaster or an economic shock, such as a sharp, unexpected rise in food and fuel prices. They also include shocks that are specific to the household such as the death or illness of a family member or the loss of a job.

Households were asked to nominate the different ways that they coped following the experience of a negative shock, in order to maintain their level of well-being. In line with much of the literature on coping with risk in developing countries, these coping mechanisms centred on: directly producing consumption items, such as food; increasing household income; adjusting or outright reducing households' consumption; and drawing down on household assets (Glewwe and Hall, 1998; Moser, 1998; Kochar, 1999; Hoddinott and Quisumbing, 2008).

Importantly, a number of the responses that were given were also directly attributable to the custom economy. Some of these responses highlighted the strength of the custom economy in allowing households to cope with shocks, including eating more food from gardens or from the reef; increasing the use of traditional support mechanisms, such as seeking additional help from family, friends and neighbours; moving in with family/*wantok*; and switching to traditional medicines. Other responses, however, demonstrated the tenuousness of the custom economy and the extent to which it is being undermined, including contributing less money at different community fundraising events; giving less money to family/*wantok;* and giving less money to the church (see Table 6.1).

In total, 86 per cent of all households surveyed indicated that they experienced a negative macroeconomic shock associated with the strong rise in international food and fuel prices and the subsequent GEC. Limiting the sample to households that experienced the shocks, 89 per cent of respondents indicated that they used at least one of the custom economy responses listed above to manage their vulnerability. Rural households were more likely than urban households to use the custom economy to deal with a shock, with 92 per cent of households, compared with 85 per cent, respectively. The near-universal use of the custom economy is a clear illustration that it remains an important dimension of Melanesian households' sense of resilience.

Looking at each of the response types in turn, by far the most prominent household response to the effects of the shocks was to turn to the natural environment, in particular the garden. Once again, there was a significant discrepancy between urban and rural households. While around 90 per cent of rural households coped with the effects of an unexpected adverse shock by increasing their use of the garden, this fell to around 61 per cent for urban households. At a community level, there was a close association between having access to a garden and using it in a time of shock. This implies that in areas where gardens are not available, households are being deprived of an important alternative source of food and income, as well as being forced to resort to other mechanisms in order to cope with shocks.

Around 57 per cent of households used traditional support mechanisms – that is sourced cash or in-kind assistance from friends/neighbours/*wantoks,* moved in with *wantok,* or switched to traditional medicines. In each case, households' in rural areas were more likely to use these mechanisms than households in urban areas. After using traditional medicines in place of modern medicines, the most popular traditional support mechanism was sourcing additional help from family, then a friend or neighbour.

Interestingly, 36 per cent of households indicated that they had jettisoned, to some extent, their customary obligations in response to a shock. Urban households were broadly similar to rural communities in this regard. Whether this is a reflection of greater targeting of contributions on the basis of need, a shift to what Holzman and Jørgensen (2000, p. 8) refer to as 'balanced reciprocity' – in which gifts are provided less on the basis of need but rather on the basis that there will be a counter gift in return – or simply a broad-based withdrawal from the system is not clear. One clue is that households appeared to clearly prioritize withdrawing their financial contributions to community fundraising events and toward family and friends than tithes provided to the church. In any case, given the threats the informal safety net already faces from rapid rates of urbanization and monetization, the ongoing ability of the traditional support systems to provide insulation from shocks should therefore be the focus of future research.

Table 6.1 Percentage of households undertaking custom economy responses to a negative shock

	Total	Urban	Rural
Used Custom Economy	89	85	92
Used traditional support mechanism	57	52	60
Used traditional medicine	38	36	40
Got additional help from family	32	29	33
Got help from friend/neighbor	19	14	22
Got additional help from church	3	2	3
Moved in with family/*wantok*	1	2	1
Used environment	84	73	90
Garden	79	61	90
Coral reef	55	49	59
Jettison Custom Economy	36	37	35
Reduced contribution at fundraising events	25	25	24
Reduced contribution at custom ceremonies	22	23	22
Reduced contribution to family/*wantok* event	18	21	19
Reduced contribution at church	8	9	7

Source: The authors.

6.5 Conclusion

The custom economy plays a significant role in the lives of most people in Melanesia. In addition to providing an important sense of identity, and a connection to the region's ancient history, the central tenets of the custom economy, including subsistence from the local environment and risk pooling, provides households with a myriad of practical benefits. These include important social protections that mitigate the effects of risk and provide an important dimension of resilience to exogenous shocks.

A unique household survey has facilitated an examination of some of the salient dimensions of the traditional economy in Melanesia. It finds that mutual support networks between extended family and *wantok* remain relatively robust even in urban localities.

While kinship networks and the custom economy do provide significant benefits to both rural and urban dwellers, it is apparent from the survey that the demands for mutual support can sometimes be difficult to meet. This appears to be particularly the case in urban areas, where the combined pressures of limited availability of land and rising costs of living mean that some urban households are finding it increasingly difficult to meet their custom obligations. Various strategies to reduce the burden of responding to social obligations and traditional exchanges were identified in the survey. It is apparent that already some households are opting out of the demands of extended family and *wantok*s, and given the experience of other Melanesian countries there is a likelihood of further erosion of the custom economy and social exchanges associated with it. The implication of this is that households and communities will increasingly lose an important form of resilience when they experience hardship. This increases their vulnerability to future shocks and could potentially lead to response mechanisms with long-term deleterious implications for human capital, such as taking children out of school and turning to crime.

Given the significant contribution being made to the sustainable well-being of Melanesians by the custom economy, it is critical that the governments of these countries include it in their national development plans and policy making. The efforts of the MNCC to rearticulate the indicators of well-being by including dimensions of the custom economy are a welcome opening gambit in this sense. There should be further research carried out on the different aspects of this economy including its environmental impacts, its potential for organic farming and its contribution to social order and political stability.

References

Abbott, D. and Pollard, S. (2004), *Hardship and Poverty in the Pacific: Strengthening Poverty Analysis and Strategies in the Pacific* (Pacific Department, Asian Development Bank: Manila).

Alderman, H and Paxson, C.H. (1994), Do the Poor Insure? *Working Paper 1008* (Agricultural Policies Division, Agriculture and Rural Development Department, World Bank: Washington).

Bedford, R. (2012), *Population Movement in the Pacific: A Perspective on Future Prospects*, Paper presented at the Conference 'Towards New Island Studies: Okinawa as an Academy Node Connecting Japan, East Asia and Oceania', University of the Ryukyus.

Belshaw, C. (1965), *Traditional Exchange and Modern Markets* (Prentice-Hall: Englewood Cliffs).

Bolton, L. (2003), *Unfolding the Moon; Enacting Women's Kastom in Vanuatu* (University of Hawai'i Press: Honolulu).

Bre, H. (2006), The Real Cost of Bride Price, *Melanesian Journal of Theology*, 22(2): 8–18.

Brookfield, H. (1973), *The Pacific in Transition: Geographical Perspective on Adaptation and Change* (Edward Arnold: London).

Chowning, M.A. (1977), *An Introduction to the Peoples and Cultures of Melanesia* (Cummings Publishing: Menlo Park).

Chung M. and Hill, D. (2002), Urban Informal Settlements in Vanuatu: Challenge for Equitable Development, *Report prepared for Pacific Islands Forum Secretariat and UN Economic and Social Commission for Asia and the Pacific*, Pacific Operation Centre, Suva.

Connell, J. (2010), Pacific Islands in the Global Economy: Paradoxes of Migration and Culture, *Singapore Journal of Tropical Geography*, 31: 115–29.

Connell, J. (2011), Elephants in the Pacific? Pacific Urbanisation and its Discontents, *Asia Pacific Viewpoint*, 52(2): 121–35.

Crocombe, R. (2001), *The South Pacific* (University of the South Pacific: Suva).

Dercon, S. (2002), Income Risk, Coping Strategies, and Safety Nets, *The World Bank Research Observer*, 17(2): 141–66.

Douglas, O. (1967), *A Solomon Island Society: Kinship and Leadership among the Siuai of Bougainville* (Beacon Press: Boston).

Duffy, F. (2008), Comparative Economic Systems, *Research Starters Business 1*, Ipswich, MA.

Feeny, S. (2010), *The Impact of the Global Economic Crisis on the Pacific Region* (Oxfam Australia: Melbourne).

Forsyth, M. (2009), *A Bird that Flies with Two Wings: Kastom and State Justice Systems in Vanuatu* (ANU E Press: Canberra).

Glewwe, P. and Hall, G. (1998), Are some groups more vulnerable to macroeconomic shocks than others?: Hypothesis tests based on panel data from Peru, *Journal of Development Economics*, 56(1): 181–206.

Haccius, J. (2009), Coercion to Conversion: Push and Pull Pressures on Custom Land in Vanuatu, *Justice for the Poor Briefing Notes*, World Bank, Washington [accessed October 2012] http://www-wds.worldbank.org/external/default/WDSContentServer/WDSP/IB/2009/07/08/000334955_20090708024457/Rendered/PDF/493220BRI0J4P010Box338947B01PUBLIC1.pdf.

Hoddinott, J. and. Quisumbing, A. (2008), Methods for Microeconometric Risk and Vulnerability Assessment, in R. Fuentes-Nieva and P.A. Seck (eds), *Risk, Shocks, and Human Development: On the Brink* (St. Martin's Press, Palgrave Macmillan: New York).

Holzmann, R. and Jørgensen S.L. (2000), Social Risk Management: A New Conceptual Framework for Social Protection and Beyond, *Social Protection Discussion Paper No. 0006*. The World Bank: Washington, DC.

Huffman, K. (2005), *Traditional Money Banks in Vanuatu Project* (Vanuatu National Cultural Council: Port Vila, Vanuatu).

Hughes, A. (1978), Solomon Islands – Economic Growth and Independence, *Pacific Perspective*, 7(1/2): 36–41

Jansen, T., Mullen, B.F., Pollard, A.A., Maemouri, R.K., Watoto, C. and Iramu, E. (2006), Solomon Islands Smallholder Agriculture Study, Volume 2 Subsistence Production, Livestock and Social Analysis, Australian Agency for International Development, Canberra

Jolly, M. (1994), *Women of the Place: Kastom, Colonialism and Gender in Vanuatu* (Harwood Academic Publishers: Chur and Reading).

Kidd, S., Samson, M., Ellis, F., Freeland, N. and Wyler, B. (2010), *Social Protection in the Pacific – A Review of its Adequacy and Role in Addressing Poverty* (Australian Agency for International Development: Canberra).

Kochar, A. (1999), Smoothing Consumption By Smoothing Income: Hours-Of-Work Responses To Idiosyncratic Agricultural Shocks In Rural India, *Review of Economics and Statistics*, 81(1): 50–61

Lamont, L. (1999), Social Relations in Rapport, M. (ed.) *The Pacific Islands: Environment and Society* (Bess Press: Honolulu).

McGregor, A.R., Bourke, M, Manley, M., Tubuna, S. and Deo, R. (2009), Pacific Islands Food Security: Situation Challenges and Opportunities, *Pacific Economic Bulletin,* 24(2): 24–42.

Maebuta, H. and Maebuta, J. (2009), Generating Livelihoods: A Study of Urban Squatter Settlements in Solomon Islands, *Pacific Economic Bulletin,* 24(3): 118–31.

MNCC (2012), Alternative Indicators of Well-being for Melanesia, Vanuatu Pilot Study report, Malvatumauri National Council of Chiefs, Port Vila, Vanuatu.

Mohanty, M. (2011), Informal Social Protection and Social Development in Pacific Island Countries: Role of NGOs and Civil Society, *Asia-Pacific Development Journal*, 18(2): 25–56.

Monsell-Davis, M. (1993), *Safety Net or Disincentive: Wantoks and Relatives in the Urban Pacific*, National Research Institute, Boroko, Port Moresby.

Monson, R. (2010), Women, State law and Land in Peri-Urban Settlements on Guadalcanal, Solomon Islands [accessed October 2012] http://www-wds.worldbank.org/external/default/WDSContentServer/WDSP/IB/2010/05/07/000020953_20100507115121/Rendered/PDF/544240Briefing1Guadalcanal1SolIslds.pdf.

Moser, C.N. (1998), The Asset Vulnerability Framework: Reassessing Urban Poverty Reduction Strategies, *World Development,* 26(1): 1–19.

Naidu, V. and Mohanty, M. (2009), Situation Analysis of Social Protection Policies, Services and Delivery Mechanisms in the Pacific: Final Report submitted to UNESCAP, Suva, Fiji.

Nanau, G.L. (2011) The Wantok System as a Socio-Economic and Political Network in Melanesia, *OMNES: The Journal of Multicultural Society*, 2(1): 31–55

Narokobi, B. (1983), *Life and Leadership in Melanesia* (University of the South Pacific and University of Papua New Guinea: Suva and Port Moresby).

Orathinkal, J. and Vansteenwegen, A. (2004), Towards Developing a Family Therapy for Melanesia, *Australian and New Zealand Journal of Family Therapy*, 25(3): 148–54

Polanyi, K. (1944), *The Great Transformation: The Political and Economic Origins of our Time* (Beacon Press: Boston, MA).

Ratuva, S. (2006), Traditional Social Protection Systems in the Pacific – Culture, Customs and Safety Nets, in ILO (ed.), *Social Protection of All Men and Women: A Sourcebook for Extending Social Security Coverage in Fiji: Options and Plans* (International Labour Organization: Suva, Fiji).

Regenvanu, R. (2009), *The Traditional Economy as the Source of Resilience in Melanesia* (Vanuatu Cultural Centre: Port Vila).

Rosser, J.B. and Rosser, M.V. (2005), The Transition between the Old and New Traditional Economies in India, *Comparative Economic Studies,* 47: 561–78.

Rosser, M., Barkley-Rosser, J. and Kramer, K.L., Jr (1999), The New Traditional Economy: A New Perspective for Comparative Economics? *International Journal of Social Economics,* 26(6): 763–78.

Shutler, R.J. and Shutler, M. (1975), *Oceanic Prehistory* (Cummings Publishing: Menlo Park).

Sillitoe, P. (2000), *Social Change in Melanesia: Development and History* (Cambridge University Press: New York).

Sillitoe, P. (2006), Why Spheres of Exchange?, *Ethnology*, 45(1): 1–23.

Storey, D. (2006), Urbanisation in the Pacific, State Society and Governance in Melanesia Project (Development Studies Programme, Massey University: New Zealand).

UNICEF (2011), *Situation monitoring Food Price Increased in the Pacific Islands,* [accessed March 2011], Available from: http://www.unicef.org/pacificislands/FINAL_SITUATION_REPORTING2.pdf.

VANGOV (2007) *Census of Agriculture 2007*, National Statistics Office, Vanuatu Government, Port Vila, Vanuatu. Accessed 18/7/13. Available at http://www.phtpacific.org/sites/default/files/surveys_dev_reports/101/files/VUT_Ag-Census-2007_final-report.pdf.

Ward, R.G. (1977), Internal Migration and Urbanisation in Papua New Guinea in R.J. May (ed.), *Change and Movement: Readings on Internal Migration in Papua New Guinea* (Australian National University Press: Canberra).

Waring, M. and Sumeo, K. (2010), *Economic Crisis and Unpaid Care Work in the Pacific*, United Nations Development Programme (UNDP), Suva, Fiji.
Wood, T. and Naidu, V. (2008), *A Slice of Paradise?: The Millennium Development Goals in the Pacific: Progress, Pitfalls and Potential Solutions* (Oceania Development Network: Apia).
World Bank (2012), *World Development Indictors Online Database* (World Bank: Washington).

Chapter 7
Vulnerability and Resilience in Melanesia: A Role for Formal Social Protection Policies?

Simon Feeny

7.1 Introduction

This book complements existing work on vulnerability, conducted at a national level, by examining the household impacts and responses to recent shocks in the Solomon Islands and Vanuatu. A core finding of the book provides credence to the assertion that these countries are among the most vulnerable in the world. Households report experiencing frequent shocks and an inability to cope with their impacts.

Similar to other developing countries, households in the Solomon Islands and Vanuatu are found to have experienced covariate (country-wide) economic shocks, with virtually all households reporting recent hikes in the prices they pay for both food and fuel. In general, households spend a large proportion of their incomes on these goods. Urban households, in particular, are highly exposed to fluctuations in food and fuel prices since they purchase a large proportion of their food from markets and local stores and are heavily reliant on buses for transport. Yet even in rural areas, where semi-subsistence lifestyles prevail, households source at least some of their food from local stores and they are reliant on boats and trucks for transport.

Relative to the price increases of food and fuel, the recent Global Economic Crisis (GEC) has had less of an impact on households in the Solomon Islands and Vanuatu. The vast majority of the population of these countries works in the informal sector. They have not, therefore, suffered the large falls in employment witnessed by other developing countries; although formal sector employment is reported to have fallen in a number of urban households. Due to a limited diaspora, the Solomon Islands and Vanuatu were also not subject to the large decline in remittances experienced by their Polynesian neighbours. Moreover, many households in these countries suffer from poor service delivery making any cuts in government spending on basic services less apparent.

In addition to these covariate shocks, households were also found to experience a number of idiosyncratic (household-specific) shocks. These shocks included natural disasters such as flooding and storm damage, as well as crop failure (often in rural areas) and theft (often in urban areas).

The nature of resilience in the Solomon Islands and Vanuatu also sets them apart from many other developing countries. Households and communities rely heavily on their natural environments during times of need. The most common household responses to a shock are to source more food from the garden and/or from the reef. Securing the future availability of these natural assets is crucial for the people of these countries.

The book documents how shocks have made it harder for households to meet their basic needs. It finds that lifestyles in Melanesian communities are changing rapidly with increasing demands for cash, in particular for school fees, basic foodstuffs and customary obligations. Concurrently, there are limited domestic opportunities for formal employment. The stress of having inadequate income can lead to broader social impacts including crime, violence, substance abuse and mental health problems.

The book has also challenged the often held belief that subsistence affluence prevails in the Solomon Islands and Vanuatu (Chapter 5). It finds that some households are living in severe poverty and large numbers of households are classified as vulnerable since they are close to the threshold at which households become poor. The book also finds that while traditional safety nets are very important in preventing destitution they do not cover everyone in the Solomon Islands and Vanuatu (Chapters 2 and 6). Moreover, they are beginning to disintegrate and are unlikely to be effective during large future covariate shocks such as price hikes and major natural disasters. These findings point to a need for a change in the policies of government and aid donors to ensure the lives of Melanesian households improve in the prospect of increasing economic shocks, continuing natural disasters and the potentially devastating impacts of climate change.

Chapter 1 identified a number of ways in which governments and international aid donors can help in reducing household vulnerability. These recommendations largely came from members of the communities that were visited during the research fieldwork. They included reducing the reliance on imported foods; improving the ability to generate an income; improving education; and reducing social tensions and fragmentation. However, an increasingly popular approach of developing country governments to address poverty and vulnerability is through the use social protection policies, with widespread recognition of their effectiveness. The remainder of this chapter is structured as follows. Section 7.2 defines social protection and examines social protection policies. Section 7.3 uses the findings from the preceding chapters of the book to make a case for implementing social protection policies in Melanesian countries such as the Solomon Islands and Vanuatu. Section 7.4 assesses how such policies should be designed while Section 7.5 provides some specific policy recommendations. Finally, Section 7.6 concludes with some areas for future research.

7.2 What is Social Protection?

Social protection can be categorized into formal social protection and informal social protection. Each is discussed in turn. Formal social protection commonly relates to a set of government policies which address risk or that provide households with support to reduce the impacts of shocks. While no single definition exists, social protection can be defined as 'all public and private initiatives that provide income or consumption transfers to the poor, protect the vulnerable against livelihood risks, and enhance the social status and rights of the marginalised; with the overall objective of reducing the economic and social vulnerability of poor, vulnerable and marginalised groups' (Devereux and Sabates-Wheeler, 2004, p.9).

Barrientos (2011) identifies three types of social protection policies: (i) social assistance; (ii) social insurance; and (iii) labour market regulations. Social assistance includes cash or in-kind transfers to individuals or households that are judged as vulnerable. They can include cash transfers, conditional cash transfers, pensions, disaster relief, housing, educational scholarships, fee waivers and subsidies. Social insurance refers to schemes that can reduce the risks associated with unemployment, old age, disability and poor health. It includes income protection schemes and health insurance. Finally, labour market regulations relate to policies that promote employment and minimum standards for workers such as minimum wages and employment protection legislation. However, social protection can, more broadly, include interventions which provide greater access to basic services such as health, education and water and sanitation (AusAID, 2012a).

Social protection policies aim to protect the poorest and most vulnerable in a society. In developing countries this will often include the disabled, single parents, children, the aged, the homeless, unemployed youth, those with no access to land, squatter settlers and the chronically ill.

While social protection policies are common in developed countries they are less widely implemented in developing countries and in Melanesian countries in particular. Baulch et al. (2008) devised a social protection index for countries in the Asia-Pacific, based on the extent of social protection policies, government spending devoted to these policies and the percentage of the poor that received benefits. While the Solomon Islands were not included in the index, Vanuatu had one of the lowest scores (together with Papua New Guinea and Tonga).

One form of social protection that does exist in the Solomon Islands and Vanuatu are their National Provident Funds (NPFs) which provide contributions to the pensions of formal sector workers. However, they exclude the unemployed and the majority of workers which operate in the informal sector and in doing so will not benefit the poorest members of society. Since most formal sector workers are male, NPFs also have a gender bias. While a number of other social protection polices exist in other Pacific countries (including child grants, disability benefits and grants to poor households), this is not the case for the Solomon Islands and Vanuatu (AusAID, 2012b). Labour standards laws (including minimum wages) exist in most countries of the Pacific although compliance rates are weak, as is government enforcement (AusAID, 2010).

The governments of the Solomon Islands and Vanuatu have made primary school education fee-free (with the support of international aid donors). It is often the case that households still make contributions to schools as well as paying for their children's transport, uniforms and school meals. Yet this recent and important policy can help explain why very few households in the Solomon Islands and Vanuatu reported removing a child from school following a shock. Ensuring that children complete their education is likely to be vital in providing them with resilience later on in life.

Informal social protection, traditional social protection and traditional safety nets can be used interchangeably. They relate to assistance provided by family, communities, religious organizations, Non-Government Organizations (NGOs) and Community Based Organizations (CBOs). The traditional or custom economy has long protected households in the Solomon Islands and Vanuatu following shocks. In the absence of many formal social protection schemes the family and community are the main sources of support for households during difficult times. This has prevented outright destitution in these countries since there is usually an entitlement to land and food, particularly in rural villages (Woodruff and Lee, 2010).

The *wantok* system (discussed in Chapter 6) keeps people of the same family, ethnic or social background linked together and ensures households help each other during times of need. This might include sharing food, assisting with the construction of a home (labour exchange) or paying for somebody else's school fees or bus fare (Monsell-Davis, 1993). Gift giving and reciprocity are other important features of the system and strong social networks should ensure that nobody in the community goes hungry. Land can also act as an important traditional safety net in the Solomon Islands and Vanuatu. The majority of land is communally owned in these countries and provides not just a source of food but also of identity (Mohanty, 2011). Having access to land also helps in meeting customary obligations.

To the extent that households often receive remittances following a shock, migration provides another type of informal social protection.[1] Households in the Solomon Islands and Vanuatu have not benefited from access to other countries' labour markets and most migration has been internal (see Chapter 4). Internal remittances of food, clothing and money all help in meeting a family's basic needs and customary obligations during difficult times. However, migration can also weaken the informal social protection system, with the migrant separated from the rest of their family. This can increase the work burden of those left at home, particularly the workload of women caring for children and maintaining the

1 See Brown et al. (2013) for a discussion of how migration and remittances can perform similar functions to formal social protection policies in the cases of Fiji and Tonga.

household. Moreover, if migration (often to urban areas) is unsuccessful, then a migrant might become reliant on food parcels sent from their home.

7.3 The Case for Formal Social Protection Schemes in the Solomon Islands and Vanuatu

Since traditional safety nets are so strong in the Solomon Islands and Vanuatu, the question arises as to whether formal social protection policies are required. After all, implementing such policies endangers replacing or substituting traditional social support structures. This chapter makes a case for greater formal social protection policies to be implemented. Its case is based on the following four arguments: (i) there are gaps in existing traditional safety nets; (ii) there is an increasing need for money to meet basic needs; (iii) traditional safety nets are weakening; and (iv) traditional safety nets are not effective at reducing vulnerability to covariate shocks. Each argument is discussed in turn.

As discussed in Chapters 2 and 6, traditional safety nets don't cover everyone in Melanesian societies. It has been recognized for some time that there are people outside of the *wantok* system with nobody to assist in times of need or that are unable to assist (Morauta, 1984; Mounsell-Davis, 1993). Chapter 5 of this book also confirms this finding with a significant proportion of households reporting that are unable to rely on anybody in the event of someone in the household getting into financial difficulties and needing support.

AusAID (2012a) also finds that there are holes in traditional safety nets with gaps in informal social protection relating to a lack of access to basic services (particularly for women and girls), clean water and adequate sanitation, a lack of economic opportunities, insufficient land and diminished land productivity. AusAID (2010) finds that informal social protection doesn't operate evenly across communities and finds rising exclusion and vulnerability among a number of specific groups including children, unemployed youth, the elderly, the disabled, single mothers and those with a long term illness including those living with HIV and AIDS.[2] These groups might not be covered by either formal or informal social protection schemes.

It is also clear that traditional safety nets are unable to prevent the incidence of poverty and vulnerability all together. Chapter 4 highlighted the issue of food insecurity for households in the Solomon Islands and Vanuatu. Some households are going hungry. It is not uncommon for adults and even children to go a day without food due to a lack of money. The majority of households reported that they worried their food would run out before they got money to buy more. Further, Chapter 5 demonstrated that there are some households living in severe poverty in the Solomon Islands and Vanuatu and that a large number of households are

2 Gibson (2006) confirms that remittances do not necessarily go to those most in need in the cases of Papua New Guinea and Samoa.

deemed vulnerable since they are close to being or becoming poor. It finds that the incidence of poverty is highest in urban and remote communities. Traditional safety nets are not ensuring that households have an adequate standard of living in these areas, contributing to the case for formal social protection policies.

Increasing monetization is also undermining the role of traditional safety nets in Melanesian countries. While meeting basic needs could often be achieved through the exchange of food and other goods, the increasing need for money to purchase food, pay for school fees and meet customary obligations means that some households are simply unable to assist others during difficult times. The fieldwork undertaken in the Solomon Islands and Vanuatu revealed that even households in very remote areas use, and therefore have a need for, money. Without money households find it difficult to meet their basic needs and income earning opportunities in these countries are limited. This can lead to social fragmentation and breakdown.

It is also argued that traditional safety nets are under strain or breaking down (ADB, 2010; Connell, 2011; Woodruff and Lee, 2010). Such arguments are again confirmed by the findings in this book. Chapter 6 found that while traditional safety nets are very important in the Solomon Islands and Vanuatu, meeting customary obligations can actually be a burden for some households, particularly given the rising cost of living. Increasing urbanization is also undermining traditional support mechanisms as social ties tend to weaken in informal urban settlements and there is a decline in reciprocity due to separation from the family (AusAID, 2012a). Further, as outlined in Chapters 2 and 4, higher rates of urbanization are leading to a move away from subsistence agriculture and an increasing reliance on imported food. Some urban households face difficulties in securing enough food due to a lack of access to land for gardens.

Finally, as put forward by Woodruff and Lee (2010) and AusAID (2010), while traditional support mechanism might be effective at reducing vulnerability from idiosyncratic shocks, affecting a small number of households, they are less effective with regard to covariate shocks which affect a whole community or country (for example natural disasters and price shocks). Shocks affecting a whole country imply there will be very few if any households with the capacity to respond and help others in need. With climate change and the increasingly integration of households in the Solomon Islands and Vanuatu into national and international economies, comes greater exposure to covariate shocks and a further need for the establishment of formal social protection schemes in these countries.

7.4 What Types of Formal Social Protection Schemes are Appropriate?

As mentioned above, there is a danger that implementing formal social protection schemes will substitute traditional support systems in Melanesian countries. This would further erode aspects of the traditional economy and increase household dependency on government; impacts which should be avoided.

In many parts of the world, Conditional Cash Transfers (CCTs) have proved an effective way of improving social indicators. CCTs relate to the direct disbursal of cash to individuals or households with the condition on them to meet certain criteria, such as sending their children to school or vaccinating them against disease.

Unfortunately there are a number of reasons to question the applicability of CCTs in Melanesian countries like the Solomon Islands and Vanuatu. Firstly, their effectiveness depends upon complementary factors such as infrastructure. For example, a CCT programme that aims to increase school attendance needs the existence of an adequately resourced school and the means for families to access the school. This might exclude very poor or remote communities without schools or roads to access them (Haider, 2010). The same applies to access to health clinics if cash transfers are conditional on vaccinations. Governments would have to be able to meet the increased demand for services that CCTs might induce. This includes ensuring that schools have teachers and clinics have nurses/physicians.

Secondly, there are concerns that CCTs can disrupt family relationships or exacerbate power struggles within communities. For instance, the funds may be 'diverted' by the household head, often the male, and cash transfers have been associated with some cases of increased domestic violence (Taylor, 2010). Thirdly, to effectively receive and make use of financial payments, households need access to financial services, which are currently lacking in many rural areas of the Solomon Islands and Vanuatu. While CCTs could be implemented in urban areas in which households have access to health, education and financial services, there is a risk that this will lead to greater urbanization.

AusAID (2012b) argues that there is little evidence to suggest that CCTs are effective due to the conditions imposed and that it is the actual cash that makes the difference. On these grounds providing poverty targeted transfers should be preferred. However, AusAID (2012a) argues against poverty targeted transfers since (i) traditional safety nets prevent hunger and destitution (in the main); (ii) they would be socially divisive through the selection of some families eligible for support at the expense of others; and (iii) might accelerate migration from poor/rural areas to informal urban settlements. While the findings from this book clearly contest the first of these points, the latter two assertions need consideration although there is currently little evidence from Melanesia to substantiate them.

Instead, AusAID (2012b) recommends universal cash transfers, such as pensions, child grants or disability payments. These are less controversial. There is evidence of pension schemes, disability benefits and a family assistance programme operating successfully elsewhere in the Pacific and by reducing dependency on others such schemes could in some circumstances, actually improve relationships with the extended family (AusAID, 2012b).

Woodruff and Lee (2010) recommend the adoption of so-called Community Based Social Protection Schemes. They acknowledge that NGOs and the churches are already active in the Solomon Islands and Vanuatu, providing health and education services to vulnerable groups and assisting people with disabilities and victims of domestic violence, for example.

Woodruff and Lee (2010) argue that using community organizations to deliver basic services overcomes the State's limited capacity to provide them and can also strengthen the organizations that deliver them. The recommend that block grants be provided to communities by the government to support vulnerable groups through the informal systems which already exist. Importantly, if these schemes build on already established informal safety nets, they can complement them and are less like to 'crowd- out' these existing forms of support than government operated schemes.

Woodruff and Lee (2010) are also aware of the challenges associated with such schemes. For example, there is the possibility of rent-seeking by village chiefs and assistance not getting to those most in need (households that already marginalized). Often the most vulnerable are those without social support and assistance from community organizations. If households do not have access to prevailing informal social safety nets they are unlikely to benefit from such schemes (which might even marginalize them further). There are also questions over whether the schemes would effectively meet the needs of women and concerns over the capacity of communities and governments to run and administer them.

7.5 Recommendations

In the absence of existing social protection policies and a very strong case for their implementation, governments need to experiment with alternative approaches. Firstly, the governments of the Solomon Islands and Vanuatu could run small pilot CCT schemes in different areas of their country and rigorously evaluate them to ascertain their effectiveness. Areas with poor health and low levels of education should be targeted but services must also be available for households to access. This will test some of the assumptions of how and why CCTs might not be appropriate in Melanesian countries. Cash grants such as pensions, child grants or disability payments are far less controversial and should also be initiated although questions remain as to who receives them and how, particularly in remote rural communities. The same applies to the community based schemes advanced by Woodruff and Lee (2010) although their relative effectiveness must be assessed against other social protection programmes.

Under a broad definition of formal social protection, governments can undertake a number of other policies which will increase household resilience in the advent of shocks. To improve levels of education, governments could consider providing free transport to and from schools and extending fee-free education to the secondary level. Providing healthy, free school meals will also encourage attendance. Encouraging the enrolment of girls is particularly important given the high prevailing rates of gender inequality in the Solomon Islands and Vanuatu.

Actions to improve access to health services, water and sanitation and financial services will all assist in building the resilience of communities.

In the design of any social protection policies, it is imperative to carefully consider any gender impacts. This requires assessing the different experiences of poverty and vulnerability of both men and women (Holmes and Jones, 2010). Such an analysis has been undertaken in Chapter 3 of this book. It demonstrated that the formal sector in Melanesia is dominated by men while women bear disproportionate responsibility for maintaining the household (including collecting water and wood, caring for children and looking after the sick and elderly). It also outlined how women in the Solomon Islands and Vanuatu bear a unequal share of the burden in adjusting to household shocks. They take on additional work (such as sourcing more food from the garden and selling produce at local markets) while continuing to meet their traditional responsibilities. During difficult times, women reported a reduction in the consumption of food and an increase in domestic violence. Successful social protection policies must therefore take into account and address these inequities.

Examples of social protection responses which address the specific vulnerabilities of women include addressing health care for women, programmes to increase education for girls and to support women's empowerment, child protection and gender based violence services, programmes to raise awareness on women's rights and public works programmes for women (AusAID, 2012c). Social protection policies should not reinforce traditional gender roles by only targeting women as mothers (Holmes and Jones, 2010).

Ultimately any choice of social protection policies must be that of the governments of the Solomon Islands and Vanuatu. However, the costs of social transfer programmes are high and it is likely they would require financial support from international donors. The introduction of pension schemes in the Solomon Islands and Vanuatu would cost approximately 1 per cent and 0.5 per cent of GDP respectively and a child support grant scheme a further 1.8 per cent and 0.7 per cent of GDP (AusAID 2012d; 2012e). The schemes can also place a large administrative burden on governments.

7.6 Areas for Future Research

In examining household vulnerability and resilience in the Solomon Islands and Vanuatu, this book has provided new insights into an under-researched issue. By analysing how households are affected by shocks and how they respond, the book has inevitably thrown up a number of further questions and is able to highlight a number of areas for future research.

Access to a garden is clearly fundamental to household resilience in the Solomon Islands and Vanuatu. The vast majority of households in the Solomon Islands and Vanuatu have a garden and sourcing more food from the garden is by far the most common household response to a shock. However, as discussed above, there are high levels of food insecurity in these countries indicating that having access to a garden is necessary but not sufficient for food security. Either gardens are not producing enough food for the household or households have become reliant on food from other sources, such as local stores. In addition, as households turn towards imported food, nutrition needs to be monitored closely. Child health is falling to unacceptable levels: 26 per cent of children under the age of five in Vanuatu and 32 per cent of children in the Solomon Islands are found to be stunted (AusAID, 2012b). There is an urgent need to further examine food security and the nutritional impacts of changing diets in these countries.

Mapping and monitoring poverty is also an important area for future research. Building upon the analysis of Chapter 5, more data needs to be collected and analysed to provide better insights in to the nature of poverty and how it varies across regions and groups within these countries. The determinants of poverty need to be explored, particularly of chronic poverty and insights are needed into how households in the Solomon Islands and Vanuatu move into and out of poverty. Systems also need to be in place to monitor the impacts of economic and other shocks. Developing effective strategies to reduce vulnerability and increase resilience are heavily reliant on such systems.

Finally, maybe the biggest question facing policymakers in the Solomon Islands and Vanuatu is how the traditional economy can be maintained and even strengthened in these countries pursuit of development. Access to land is integral to households and communities in Melanesian countries. AusAID (2012a) argues against land reform such as releasing land for commercial purposes since this is likely to lead to a further degeneration of traditional safety nets. Those living on custom land, that participate in traditional ceremonial activities and who are active members of their community are, on average, found to be happier (MNCC, 2012). Social support networks are also clearly important to households and communities and should be a very important consideration in the pursuit of development. Lifestyles are changing fast in Melanesia and more research is needed to help inform and shape these countries' future.

References

ADB (2010), *Weaving Social Safety Nets* (Asian Development Bank: Manila).

AusAID (2010), Social Protection in the Pacific – A Review of its Adequacy and Role in Addressing Poverty (Australian Agency for International Development: Canberra).

AusAID (2012a), Informal Social Protection in Pacific Island Countries – Strengths and Weaknesses, AusAID Pacific Social Protection Series: Poverty, Vulnerability and Social Protection in the Pacific (Australian Agency for International Development: Canberra).

AusAID (2012b), Achieving Education and Health Outcomes in Pacific Island Countries – Is There a Role for Social Transfers? AusAID Pacific Social

Protection Series: Poverty, Vulnerability and Social Protection in the Pacific (Australian Agency for International Development: Canberra).

AusAID (2012c), Social Protection and Gender – A Life-cycle Approach, AusAID Pacific Social Protection Series: Poverty, Vulnerability and Social Protection in the Pacific (Australian Agency for International Development: Canberra).

AusAID (2012d), Poverty, Vulnerability and Social Protection in the Pacific: The Role of Social Transfers, AusAID Pacific Social Protection Series: Poverty, Vulnerability and Social Protection in the Pacific (Australian Agency for International Development: Canberra).

AusAID (2012e), Micro-Simulation Analysis of Social Protection Interventions in Pacific Island Countries, AusAID Pacific Social Protection Series: Poverty, Vulnerability and Social Protection in the Pacific (Australian Agency for International Development: Canberra).

Barrientos, A. (2011), Social Protection and Poverty, *International Journal of Social Welfare*, 20: 240–49.

Baulch B., Weber, A. and Wood, J. (2008), *Social Protection Index for Committed Poverty Reduction, Vol. 2* (Asian Development Bank: Manila).

Brown, R.P.C., Connell, J. and Jimenez-Soto, E.V. (2013), Migrants' Remittances, Poverty and Social Protection in the South Pacific: Fiji and Tonga, *Population, Space and Place*, [forthcoming].

Connell, J. (2011), Elephants in the Pacific? Pacific Urbanisation and its Discontents, *Asia Pacific Viewpoint*, 52(2): 121–35.

Devereux, S. and Sabates-Wheeler, R. (2004), Transformative Social Protection, IDS Working Paper 232, Institute for Development Studies, Brighton. http://www.ids.ac.uk/UserFiles/File/poverty_team/social_protection/wp232.pdf [accessed September 2012].

Gibson, J. (2006), Are There Holes in the Safety Net? Remittances and Inter-household Transfers in the Pacific Island Economies, Working Paper Number 1, Pasifika Interactions Project.

Haider, H. (2010). Access to Quality Education Program, Fiji: Framework for Delivery (AusAID: Canberra). [accessed June 2012], http://www.ausaid.gov.au/countries/pacific/fiji/Documents/fiji-aqe-design-07-feb-2011.pdf.

Holmes, R. and Jones, N. (2010), Rethinking Social Protection From a Gender Lens, Overseas Development Institute Research Paper, London.

MNCC (2012) Alternative Indicators of Well-being for Melanesia: Vanuatu Pilot Study Report, Malvatumauri National Council of Chiefs, Port Vila, Vanuatu

Mohanty, M. (2011), Informal Social Protection and Social Development in Pacific Island Countries: Role of NGOs and Civil Society, *Asia-Pacific Development Journal*, 18(2): 25–56.

Morauta, L. (1984), Left Behind in the Village, Institute of Applied Social and Economic Research Monograph No.25, Port Moresby, Papua New Guinea.

Mounsell-Davis, M. (1993), Safety Net or Disincentive: Wantoks and Relatives in the Urban Pacific, National Research Institute Discussion Paper No.72, Port Moresby, Papua New Guinea.

Taylor, S. (2010), Cash transfers cannot revolutionise aid: Giving money direct to poor families is a valuable idea, but only as part of a wider system of co-ordinated donation, in *The Guardian*. [accessed April 2013], http://www.guardian.co.uk/commentisfree/2010/jul/02/cash-transfers-international-aid-development.

Woodruff, A. and Lee, S. (2010), Social Protection of the Vulnerable in the Pacific, in S.W. Handayani (ed.), *Enhancing Social Protection in Asia and the Pacific: The Proceedings of the Regional Workshop* (Asian Development Bank: Manila).

Index

Bold page numbers denote figures; italic page numbers denote tables.
n denotes footnotes

For Product Safety Concerns and Information please contact our
EU representative GPSR@taylorandfrancis.com Taylor & Francis
Verlag GmbH, Kaufingerstraße 24, 80331 München, Germany